Powder X-ray Diffractometry in the Analysis of Materials
Utilization of MiniFlex

Jimpei HARADA

MARUZEN PLANET

Powder X-ray Diffractometry in the Analysis of Materials
— Utilization of MiniFlex —

The English translation published by Maruzen Planet Co., Ltd., Tokyo.

printed in JAPAN

Preface

X-rays irradiated onto a material are scattered back along several specific directions, having a fairly sharp profile. This phenomenon is known as Bragg reflection of X-rays by crystallites in the material. When the intensities of scattered X-rays are plotted against the scattering angle, a graph characterizing the material is obtained. This graph is called powder X-ray diffraction pattern to that material. If you know a basic idea to analyze the pattern, you are able to confirm the structure of crystallites. Thus, even if you encountered an unknown material you may identify the structure of crystallites existing in the material by observing diffraction pattern. This is structure analysis of unknown material by X-ray diffraction method.

Since the photographic film was known to be sensitive to X-rays, radiographic films being more sensitive to X-ray or radiations were especially developed and had been used to observe X-rays. Several X-ray diffraction cameras based on such radiographic films had been designed. Among them a powder X-ray diffraction camera had been known well. This camera was of very convenient to use. In observing diffraction pattern the amount of sample was something like that of sewing needle. Thus, this camera had brought a great power in the field of material science and technology. It had been utilized in not only educations but also research studies in physics, chemistry, metallurgy, mineralogy, and material engineering in university.

The X-ray powder diffraction camera is convenient and easy to handle, as mentioned, but has a demerit also. We could not say that the radiographic films are of sensitive to X-rays, as we need very long exposure time. Besides, it is time-consuming work, because the X-ray image does not appear until the film exposed is sufficiently developed in a darkroom. X-rays diffractometer of the convergent optics by using the detector, which has fairly high efficiency compared with X-ray film, appeared out of nowhere as if to make up such demerits of diffraction camera. It was the diffractometer which was devised in such a way that all the X-rays diffracted from a wide region of powdered samples, which are scattered on a board, could be converge at the receiving slit in front of the detector. Thus, the measurement of diffracted X-rays was very efficient with high resolution and the data obtained in this way could be automatically recorded in recording paper. Because a recording medium became to be able to store the data, the comparison with existing data became easy. The usefulness of the device was recognized by every users especially for their research studies in comparison with the X-ray powder diffraction camera and it was continued up to the present day. However, unfortunately the devices had not spread out for training personnel that much. May be, the reason was simply in the fact that such electronic system of detector, amplifier, scaling and controller units was fairly expensive around that time.

More than twenty years ago, Rigaku Corporation marketed a desk type X-ray diffractometer based on convergent optics. The desk type means that the device is small enough so that it is possible to put it on any desk in research laboratory. The X-ray diffractometer may be said even now to be the smallest one in the world. It was called simply MiniFlex as its brand name. Because it has been more than twenty years from that time, the early model has been improved now so as to be able to equip particularly with one-dimensional detector that is Rigaku's proud in its performance. It is called MiniFlex 300/600. Such an improvement would be hard to imagine before thedecade. Besides high temperature and also cryogenic sample stages became available to use even for this small diffractometer.

Under such circumstances, Mr. Hikaru Shimura, the president of Rigaku Corporation, asked me to write such a guide book which familiarizes beginners with X-ray diffraction phenomenon by reading it with utilizing Rigaku's MiniFlex. His requirement was indeed nice word to me, but when I got started writing it, a following reality came across in my mind. Most of X-ray diffractometers nowadays are designed to be able to get the measurements automatically accomplished if you put your sample on the sample stage of the device and follow the instruction displayed on the screen of the control computer. Furthermore, becausc any existing software is available to use in the analysis, a general answer will be given without any special knowledge. This is a nowadays trend, telling that it is unnecessary for user to know the detail how to analyze the data obtained. On the other hand, the next proposition comes out also. Is it acceptable or not for us to entrust all of them to a computer?

There is a big difference among two generations when modern times is compared with the days I received education. The diffractometer, which is controlled by computer, and the software of structure analysis did not exist at all on the olden days. My first experimental study was carried out by the direct guidance of a senior research associate, using text books listed on the end of this book. It means that I my self was rather brought up in this way. On the basis of the knowledge obtained by those experiences I could continue own studies by paying attention to the new subjects in the field of materials science.

If returning to the present subject, it is found that for the beginner who wishes to understand materials science, there is no way to rely on other than the instruction of the computer and the reference book of high level. Because it is not easy for beginner to come across a good senior adviser at this time, it was understood that he needs rather a gentle guide book. This is a reason that I made up my mind to write this guide book. I started to write this book for the beginner who wishes to become able to make skillfully use of MiniFlex 300/600 as an analytical equipment and also to understand the structure of crystallites in material on an atomic scale by reading this book.

Generally, it is true that the X-ray diffraction technology including the data analysis is based on the existing kinematic diffraction theory. When a new unknown material coming out, it happens that a new experimental method has to be considered. Inversely, when a new technology being introduced, we can make sure that a new knowledge about material is given. We can say that diffraction crystallography has evolved by stepping such steps. Researcher, even when reliable X-ray diffraction data were obtained, always, should be in an attitude to contemplate if there is a problem or not in the data obtained. That point was kept in mind so that you can learn through this booklet. I wish that this textbook is used as basic one for beginner so as to be able to read a more professional advanced textbook as a next step.

I took unexpectedly number of years until the completion of this book. Nevertheless Mr. H. Shimura, the President of Rigaku Corporation, and Mr. K. Suzuki, General Manager of Rigaku Corporation, have encouraged me constantly from beginning to end. I express here my heartfelt thanks to them, and also to Dr. Ken-ichi Ohshima, Professor Emeritus of Tsukuba University, who has always encouraged me in my completing this book.

All of the experimental data presented in this book are what have been measured kindly by skillful three engineers of Rigaku Corporation: Mr. A. Dosho, Ms. Y. Namatame and Ms. Y. Iwata, by using MiniFlex 300/600. I had useful comment also from Dr. T. Taguchi in writing semiconductor detectors in chapter 6. I wish to thank to those staff member of Rigaku Corporation.

This book is what has been translated into English. The original book written in Japanese was published in October 2015 by Maruzen Pub. Co. Both the English and Japanese versions are based on the financial support of Rigaku Corporation. In editing the English version, I was specially taken care by Mr. M. Nobe and Mr. T. Konishi of editorial planning group of Maruzen Planet Co. Ltd. There were also several fine works such as the remake of tables and figures for English version, those were done carefully by Mr. Tetsuya Ozawa of Rigaku Corporation and Mr. Takeshi Osakabe. I like to thank to all of those people. Upon translation, Dr. Jungeun Kim, who was research scientist of SPring-8, helped me in many respect with translation, by pointing out the mistakes during the reading carefully the Japanese version of the first edition. If she was not, I think that it was not able to see the completion of the English version of such a form. I thank in respectfully to her.

August 2016 Jimpei Harada

Contents

Part 1
X-ray material analysis

How much of knowledge about X-ray diffraction do we need when we would like to carry out sufficient analysis of a given material? This is an inquiry which I am often asked. It is natural to have it, because it requires to know well the method of analysis called the X-rays diffraction. Besides, it is hoped to have no careless or common oversight in the interpretation of provided data. It means that you need also to have enough basic knowledge about materials aimed for analysis to you.

Two scientific fields, X-ray diffraction method and materials science, were in the independent field, but there exists a historical background that provided substantial synergy. X-ray diffraction phenomenon by a crystal was one of the studies of physics. Since most of the material is consisted of the crystallites, it is also an important research target in the material science to understand the structure of the crystallites on an atomic level. So, both the field have been accompanied by the development of the necessary technology.

X-ray diffraction method is useful not only for the structural analysis of crystallite, but also for stress analysis of the material, because it is possible to measure the distortion of the crystallites constituting the material. Furthermore, analysis of texture of the crystallites present in a material, the size distribution of the crystallites, and differences in subtle physical properties of several materials are understandable now by using X-ray diffraction. Thus, it is clear that X-ray diffraction method is important tool even in the field of material science.

In this book, especially not dividing into those two scientific areas, those two are written in the first part as the title of the "X-ray material analysis". Instead, since the experimental tool to obtain the data is important science and technology necessary, those topics are taken up the "setting up of basic experimental tools" as the second part. It is so as to explain the essential technical issues related to X-ray diffraction apparatus.

Chapter 1

Trial use of MiniFlex 300/600

When an X-ray beam is irradiated to a material, the X-rays are strongly reflected from the irradiated region along several specific directions. This is due to the fact that the X-rays are diffracted by crystallites, which are the structural elements of a material. This phenomenon is referred to as Bragg reflection. X-ray diffractometer is a device to measure the intensity of the diffracted X-rays as a function of the scattering angle as accurately as possible. The graph showing the intensity of the diffracted X-rays against the scattering angle is X-ray diffraction pattern. By the analysis of the diffraction pattern, the structure of the crystallites in the material can be confirmed. The instrument that provides such a diffraction pattern is **X-ray diffractometer**. As mentioned in the preface, MiniFlex is a sub-miniature X-ray diffractometer. The Bragg reflection is usually observed not to be a simple sharp peak but with a finite width. By the analysis of those data, various type of information can be given about the composition of the material on an atomic scale. Samples that are fabricated under slightly different conditions could happen to be differentiated by the X-ray diffraction patterns obtained from them.

In the first chapter, the text book has been attempted so as to be able for the reader to understand easily the principles how to operate the X-ray diffractometer and then how to analyze the data in order to find out the structure of material on an atomic level. A proverb goes that one picture is worth a thousand words. Let us start by presenting some typical examples.

1.1 Operation principle of MiniFlex 300/600

Figure 1.1 (a) shows an explanation of the principle how to measure X-rays scattered[1)] from powder sample mounted as a flat plate. In the figure, X is a point-like X-ray source from which a parallel X-ray beam is taken out. The X-ray beam hits point O of the sample at an angle θ with respect to the sample surface. Then, the X-rays scattered come out from the O point to various directions. The detector D is set so as to be able to measure only OD X-rays which are scattered along the symmetric direction with respect to the incident X-ray beam XO. In this geometry, OD is simply making 2θ angle against the incident beam direction XO. So the angle 2θ is called the scattering angle. A measuring equipment can be said to be sufficient X-ray diffractometer, if it is designed as that the X-rays scattered from the point O on the sample can be measured with the scattering angle 2θ from 0 to 180 degree under the symmetric reflection condition.[2)]

Here, let us figure out the role of an X-ray tube which is currently used. The X-rays are emitted from a metal plate placed in a vacuum tube which is called as the X-ray tube. In the tube, electrons are accelerated and strike the metal plate by an applied voltage of the order of 10 kV to the plate. The shape of the emitter of X-rays on the target is a line segment so that it looks a line segment-like and also a spot like, depending on the directions to look at the emitter. In the use of the X-rays extracted by such an X-ray tube, we say the side of the X-ray tube to be able to see spot-like as the direction of the point focus and the other side to be able to see line segment as the direction of line focus, respectively. Thus, it is possible to select one of the two; point or line focus. The metal plate which is collided by electrons is called the target. As the target, Cr, Fe, Co, Cu, and Mo

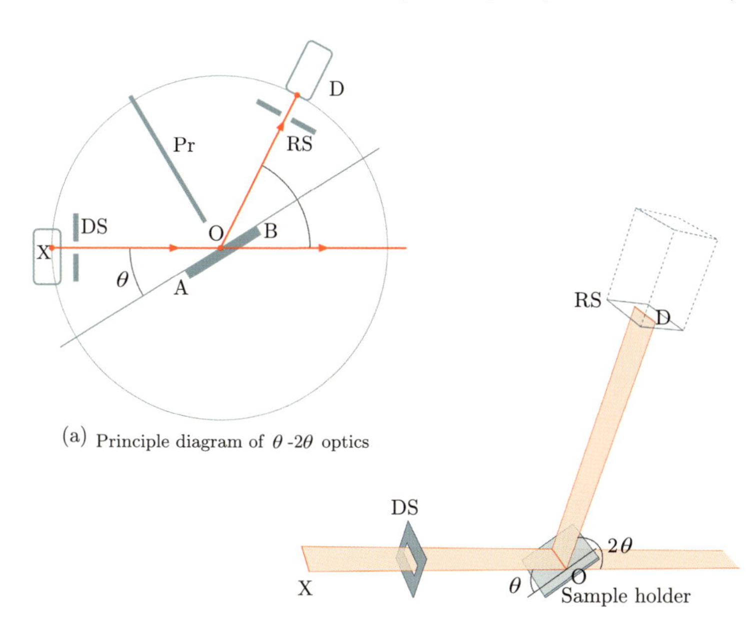

(a) Principle diagram of θ -2θ optics

(b) Optics by using line focus X-ray source

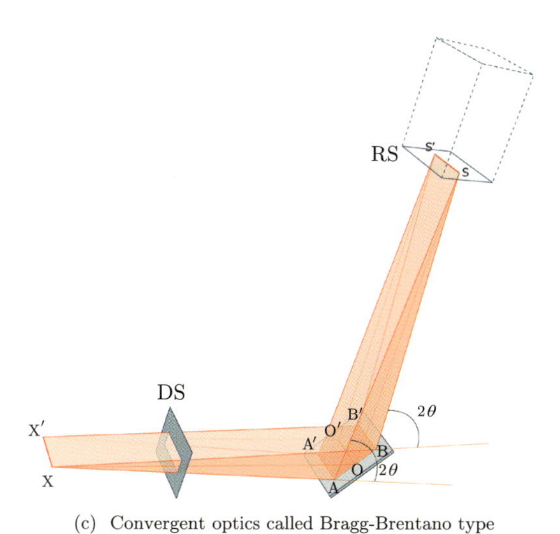

(c) Convergent optics called Bragg-Brentano type

Figure 1.1 Measurement principle of X-ray diffractometer.

and so on are used. Usually it is abbreviated as Cu or Mo tube or Cu or Mo X-ray tube. These tubes have difference in their spectra. The details are given in the Chapter 5.

Based on the principle of the measurement shown in the Figure 1.1 (a), a very narrow parallel beam formed by using a point focus X-ray tube should be made so as to irradiate a point on the sample. A narrow parallel beam can be formed by a DS (**divergence slit**) of a pinhole type which is set on a slightly distant position from the point source. The parallelity of the beam is dependent on the size of the source and slit, and the distance between the two. The more parallel the beam is, less the intensity of the X-rays is given. It is difficult to have high intensity data by using the high parallel beam. Therefore, the position of the slit from a source is an important factor to estimate the divergence angle.

As a next step, we have to set up the X-ray beam to irradiate a point O of the powder sample spread on the plate-like sample holder. The angle θ between X-ray beam and the surface of sample is the **incident angle**[3)] in the diffraction crystallography. Although X-rays are scattered in the various directions from a point O on the sample, only X-rays scattered which is symmetric with respect to the incident beam are collected by the counter. The parallel X-ray beam scattered from the O point has to be obtained. With the same technique used in making the incident X-rays, one more slit can be set in front of the detector, as shown in the Figure 1.1 (a). The pinpole slit is located at an inverse direction θ to the surface of the sample from the point O. This is **the receiving slit** and is abbreviated to **RS**. The detector (D: counter) is placed behind the RS; the X-rays scattered along that direction can be counted. The direction of the scattered X-rays, that are caught by the counter, makes an angle 2θ with the direction of the incident X-rays. This is **the scattering angle**. Besides, the plane X-O-D which are formed by the incident X-rays, the O point and the scattered X-rays is called **the scattering plane**.

In order to change the incident angle θ, there are two techniques. One of them is the clockwise θ-rotation of the arm of incident beam X-DS-O around the center of the point O by keeping the sample plane fixed. At the same time, the detector arm of O-RS-D should be rotated counterclockwise around the same center axis by the amount of θ. This is called $\theta-\theta$ rotation without the movement of the sample stage. The other one is the counterclockwise θ-rotation of the sample plane, and at the same time the counterclockwise 2θ-rotation of the detector arm of O-RS-D; this is called $\theta-2\theta$ rotation. In this technique, the incident beam can be fixed. This latter technique is usually adopted, because the instrument is stable due to no movement of heavy X-ray tube. A symmetric reflection condition with respect to the surface of the sample is also satisfied. In addition, the scattering angle can be set to any angle between 0 degree and close to 180 degree.

Such a $\theta-2\theta$ type diffractometer to which MiniFlex 300/600 belongs has been used. However, the intensity of X-rays scattered from one point O on the sample is not sufficient enough. Thus, an idea has been considered to improve so as to get X-rays scattered from more larger area of the sample.

First step is to use line focus X-ray source, as shown in Figure 1.1 (b). By using such an optics, X-rays scattered from a segment O-O' on the sample can be counted instead from one point O. Intensity observed tends to be estimated six to seven times more. However, since the intensity of the X-ray at each point of the line focus is reduced to $1/6 \sim 1/7$ compared to the case of point focus, but the amount of sample contributing to the diffraction will increase. As the result virtually intensity is not changed by this optics.

Powerful improvement is really obtained by the use of divergent X-rays instead of parallel beam. This optics is shown in Figure 1.1 (c). By using a wide divergent slit to DS, divergent X-ray beam can be formed and irradiated to a wide region of the sample. Among the X-rays scattered from each point on the sample those scattered by 2θ angle converge especially on the receiving slit RS in front of the detector. Thus, all the X-rays that came through RS are collected by the detector. Since the detector can collect X-rays scattered from large area of the sample by using this divergent beam, it is possible to make the intensity to several ten times. The details about its resolution are explained by part 2 "Setting up of basic experimental tools". In this text book, this improvement of optics has been explained by using a word "the use of the divergent X-ray beam", but it is usually called **convergent beam method**. In Japan, it is commonly called **Bragg-Brentano** optics[1].

Finally, a role of the screen, (a kind of silt) is explained, which is drawn by a thick vertical line (Pr) in Figure 1.1 (a). This screen is set perpendicularly to both the sample surface and the scattering plane. Its important role is paying in protecting the detector to collect useless X-rays scattered diffusely on the side of the incident beam. Although this screen is not drawn in Figure 1.1 (b) and (c), it is attached in the most of diffractometers.

1.2 Measurement of samples

In this section it will be shown what kind of data can be obtained by the MiniFlex 300/600 by selecting three typical materials, CsCl, α-Fe, and Al. The powder sample of CsCl, α-Fe, and Al have been set on the sample stage of MiniFlex 300/600. The results obtained are shown in Figure 1.2 (a), (b), and (c), respectively, where the intensity (vertical axis: cps; counts per sec) detected by the detector is usually plotted against the scattering angle 2θ (horizontal axes: degree). The vertical axis can simply be used "counts" which are total counts detected during a fixed time. There is another presentation for vertical axes: expressed as circumstances on the log scale. In the present measurement, the step width has been fixed to be $\Delta 2\theta = 0.02°$ and the measurement time 1.2 seconds.

The experimental condition of this measurement is as follows: a copper X-ray tube is used with the tube voltage of 30 kV and a current of 15 mA. The DS and RS slit width is 1.25° and 0.3 mm, respectively. The scintillation counter (SC) is used, but you may use the (high speed) semiconductor 1-dimensional detector, D/teX Ultra[4)], if you wish. In order to obtain high quality diffraction data with low background, the characteristic CuKα radiation is selected by setting a **highly orientated pyrolytic graphite** (HOPG[5)]) as a **monochromator** in front of the SC. (more detail : see Section 7)

Comparing the three results, a most common point is that all the data are consisted of a lot of sharp peaks at different scattering angle $2\theta_j (j = 1, 2, 3 \ldots)$. The numbers of peaks are also different from one material to another. These peaks are nothing but the **X-ray Bragg reflections** from the crystallites which are the elements that make up the sample materials. The author assume that the reader of this textbook have already been learned about X-ray Bragg reflection in class of physics of High School, but its detail description is shown in Chapter 1.4.1. The structure of the crystallites is known to be quite different, depending on materials. Consequently, all the diffraction angles and the intensities of Bragg reflections are different by the sample materials as seen from the three graphs in the Figure 1.2. These graphs are called as **diffraction pattern** in which intensity is plotted against scattering angle 2θ.

It is a matter of course that three diffraction patterns observed are not overlapped. It means that one can easily differentiate those three materials by comparing simply their diffraction patters. Once you memorize the diffraction pattern from a particular material you may compare it with other obtained from unknown sample. As the result, you can tell whether the material is the same or not. Extending this idea to many existing materials, you may have a data base for your use. However, it is not necessary to make their own database, as database worthy of trust already exists. That is referred to as ICDD, (International Centre for Diffraction Data). Comparing X-ray diffraction pattern you obtained from an unknown sample with ICDD database, you may identify the unknown material by X-ray diffraction. This is the identification of unknown materials by X-ray diffraction. You can say that it is qualitative analysis by X-ray diffraction. In the Figure 1.2 (a) – (c), the measured data are given with red color and the ICDD result with blue stick. You will see how much well the observed values agree with the values given by ICDD database.

The range of the scattering angle in above measurement is from 20° up to 120°. Within this range, a lot of Bragg reflections are observed in CsCl, but a very limited number in α-Fe and Al: 19 reflections for CsCl, and 5 and 8 reflections for α-Fe and Al, respectively. Its reason is due to the difference in their crystal structures, since crystallites of each sample have

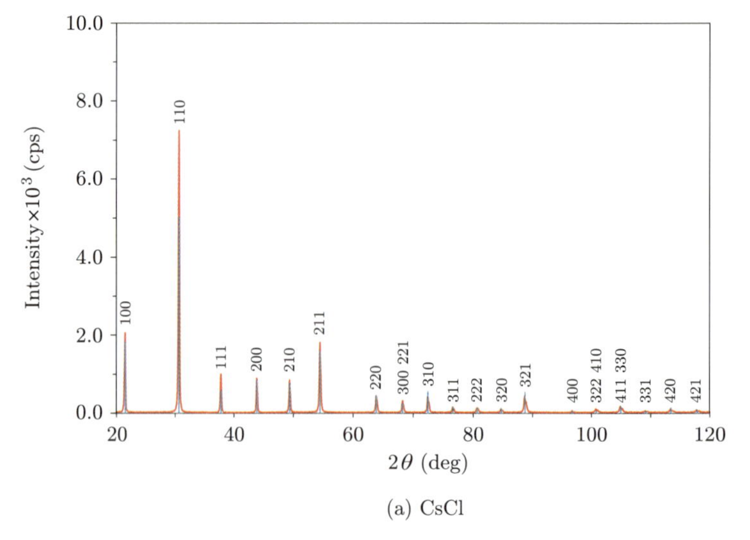

(a) CsCl

Figure 1.2 Powder X-ray diffraction pattern measured by CuKα line.

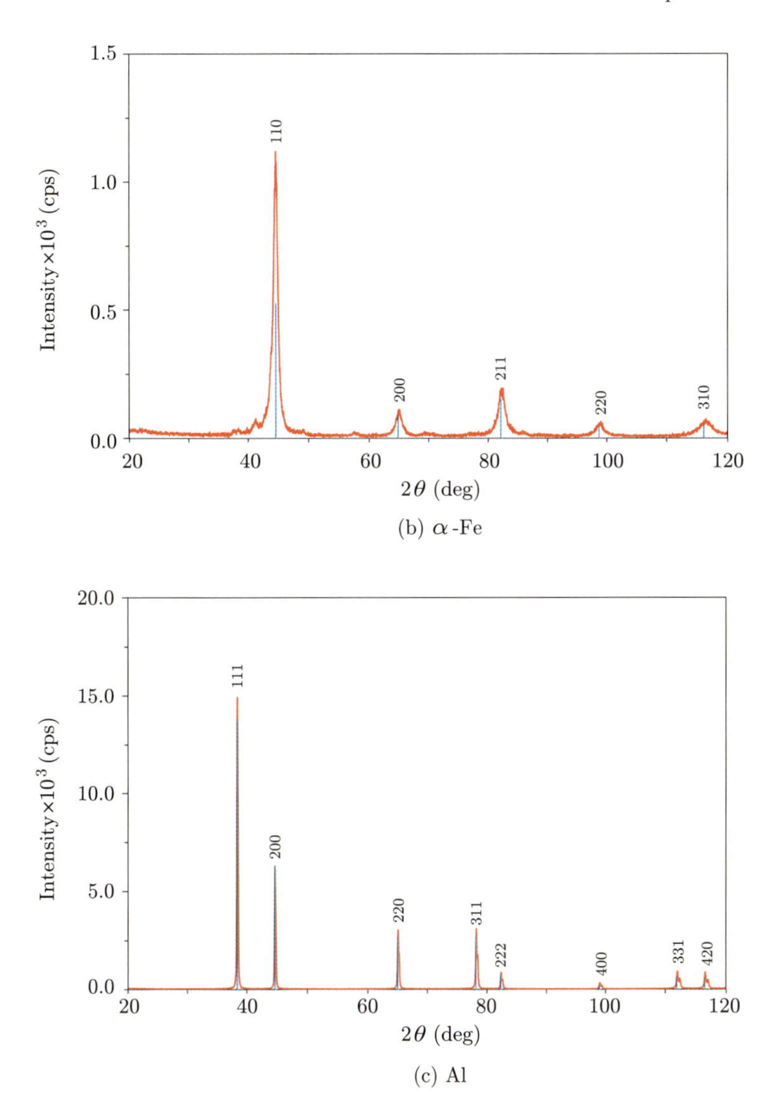

(b) α-Fe

(c) Al

Figure 1.2 Powder X-ray diffraction pattern measured by CuKα line.

their own crystal structure. As the results you will see that there are a definite number of Bragg reflections in a given range of scattering angle.

Figure 1.3 (a), (b), and (c) are the enlargement of particular Bragg reflections seen in the Figure 1.2 (a), (b), and (c), respectively. Our attention is on the shape of the Bragg reflection, which is not so simple so that it is often referred to as profile. Roughly speaking, all the profiles seen in Figure 1.3 (a), (b), and (c) are asymmetric, showing tail to the low angle side of the Bragg reflection. In the present state of affairs, ignoring the asymmetry but paying attention to the widths of their profiles, we can see that all the widths are different one another.

Most of industrial materials can be said to be in the state of aggregation of a lot of tiny crystals, which are called crystallites or crystalline particles. Furthermore, different mechanical properties of industrial materials can be said to come from the difference in the state of aggregation of crystallites. In order to differentiate the different state of aggregation, we use a technical term "crystalline textures". Really, the profile of a Bragg reflection with which we are concerned reflects the crystalline texture of the sample

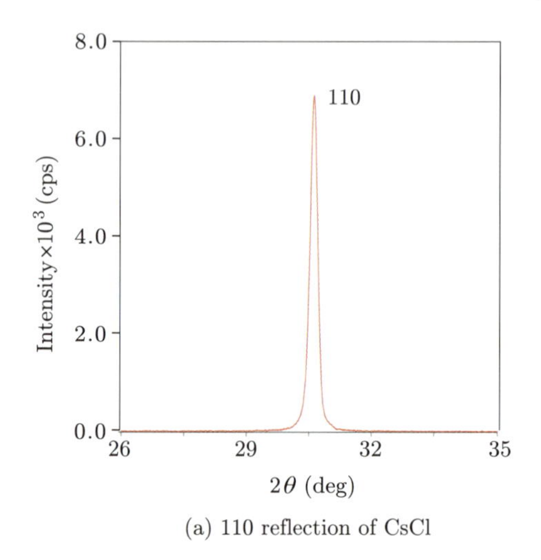

(a) 110 reflection of CsCl

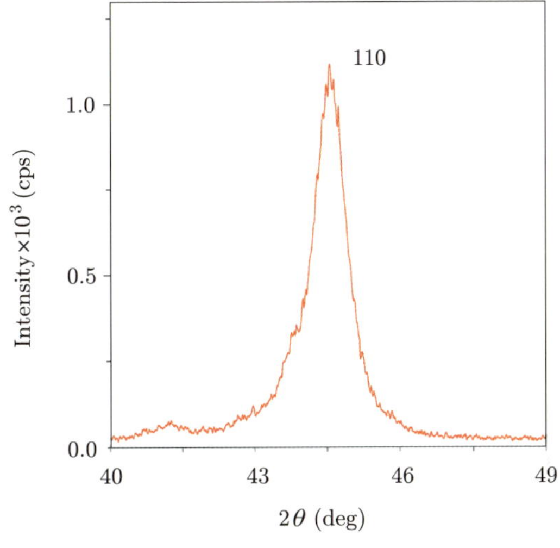

(b) 110 reflection of α-Fe

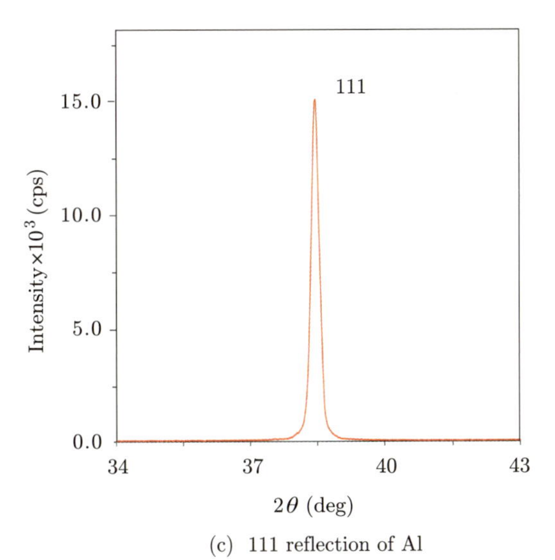

(c) 111 reflection of Al

Figure 1.3 Profile of the Bragg reflection.

material

Although it might be too simple to say, the broadening of the Bragg reflection happens when the average size of the crystallites become small. On the contrary, the Bragg reflection becomes sharp by bigger size of crystallites. A more precise analysis gives information about the size distribution of crystallites in the sample. In other words, the information of width of profile gives us a clue to understand not only the type of crystallites but also their size distribution in the sample.

Very recently a great deal of development has been progressed especially in the field of the profile analysis. It is, thus, becoming to be possible to discuss crystalline texture with considerable reliability. Accordingly, MiniFlex 300/600 has been improved also in a way that it can be utilized in analyzing data in this field. In this book, an attempt has been made so that the reader may understand the simple physics related to the scattering and diffraction phenomena and also line broadening due to crystalline texture of a given material by using X-rays.

1.3 Structure of crystal

In this section, the experimental result of Figure 1.2 is confirmed by depicting the crystal structure of CsCl, α-Fe, and Al. This is the first step to study crystallography. The geometry about crystal lattice is summarized by the title [Geometry of crystal lattice] in appendix B, so that please refer to it.

1.3.1 Unit cell

Figure 1.4 (a), (b), and (c) show the minimum basic units, which is called **unit cell**, representing crystal structures of CsCl, α-Fe, and Al, respectively. Obvious differences exist among them. First of all, the sizes of unit cells and also location of atoms are different, although three unit cells are

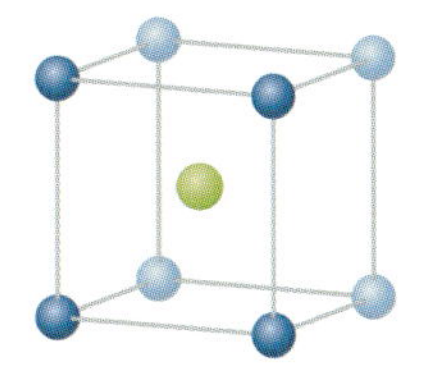

(a) Cs: 0, 0, 0
Cl: 1/2, 1/2, 1/2

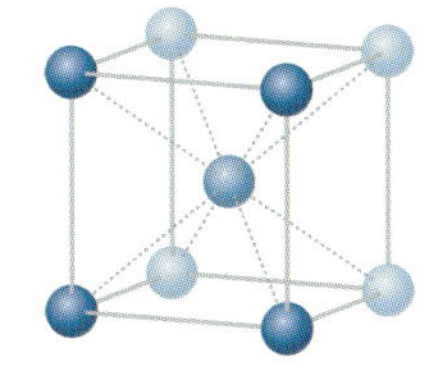

(b) Fe: 0, 0, 0
Fe: 1/2, 1/2, 1/2

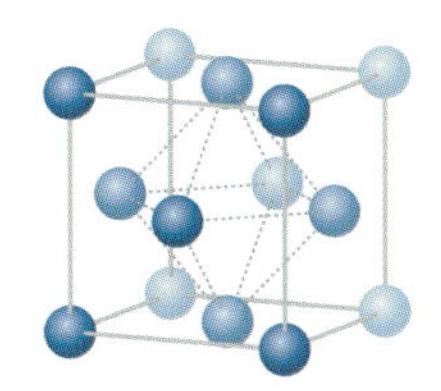

(c) Al: 0, 0, 0 0, 1/2, 1/2
Al: 1/2, 0, 1/2 1/2, 1/2, 0

Figure 1.4 Unit cell of CsCl, α-Fe, and Al crystal.

in the same cubic form. In crystallography, it is said that crystals of which unit cell is in a form of cubic are the crystals belonging to **cubic system**. In general a given unit cell is identified mathematically in terms of three vectors, $\boldsymbol{a}$, $\boldsymbol{b}$, and $\boldsymbol{c}$ which represent the three edges of the parallelepiped. The most common way to represent a unit cell is, however, to use 6 parameters that are scaler quantities: namely they are three angles α, β and γ in addition to the three lattice parameters a, b, and c, where α is the angle between vector $\boldsymbol{b}$ and vector $\boldsymbol{c}$. β is the angle between $\boldsymbol{c}$ and $\boldsymbol{a}$ and γ is the angle between $\boldsymbol{a}$ and $\boldsymbol{b}$. In the present case all the crystals belong to the cubic system so that only one lattice parameter is sufficient to represent the unit cell, because $\boldsymbol{a}$, $\boldsymbol{b}$, and $\boldsymbol{c}$ are the vectors of the same magnitude and furthermore those three vectors are perpendicular to one another. Thus, we have $|\boldsymbol{a}| = |\boldsymbol{b}| = |\boldsymbol{c}| = a_0$ with $\alpha = \beta = \gamma = \pi/2$

The atom position is usually represented by the three coordinates (x_j, y_j, z_j) of which units are taken to be the lengths along the three edges of the unit cell. The coordinates representing atomic positions are given under each Figure 1.4 (a), (b), and (c) each. As for CsCl, if the Cs ion is at the position of (0, 0, 0), Cl ion is laid on the body centered position of (1/2, 1/2, 1/2) and *vice versa*. The structure of which unit cell is given by (a) is called CsCl type-structure and many materials are of such a structure, (see Table C.3). Although the unit cell is formed by two atoms, one CsCl molecule is regarded to be occupied in the unit cell. The Figure 1.4 (b) is the structure of α-Fe. Two Fe atoms are located at the origin (0, 0, 0) and also the body centered position (1/2, 1/2, 1/2). The same kinds of atoms occupy both the positions. This is a difference from the CsCl structure. This structure is called **body centered cubic structure**. The number of atom in the unit cell is 2. Contrary to (a) and (b), the unit cell structure of Al seen in (c) is very much different. Four Al atoms are located each at (0, 0, 0), (1/2, 1/2, 0), (0, 1/2, 1/2) and (1/2, 0, 1/2) positions in the unit cell. The three atoms are at the face centered positions except one at the origin so that the structure given by this Al is called **face centered cubic structure**.

1.3.2 Miller indices

Figure 1.5 (a) shows a part of crystal which is periodically arranged with cubic structure along the x-, y- and z-axes. We consider a possible lattice plane using 3-dimensional lattice formed by cubic structure. Several lattice plane shows in Figure 1.5 (b). Each figure is explained with a matrix number: 1.1) in the figure is a top figure of the left side, and 2.3) is the position of second column and third row.

1.1) of Figure 1.5 (b) : we consider a set of parallel planes with interplanar spacing a_0 which is the lattice constant of $\boldsymbol{a}$-axis. These parallel planes are called (1 0 0) plane. A perpendicular direction of the plane is called [1 0 0] direction. The round and the square bracket indicates the lattice plane and the direction, respectively. If a plane intersects $\boldsymbol{b}$- (or $\boldsymbol{c}$-) axis, the lattice plane is written as (0 1 0) (or (0 0 1)). The direction is [0 1 0] (or [0 0 1]). With the same method, the lattice plane of $-\boldsymbol{a}$-axis is ($\bar{1}$ 0 0). We can also understand the lattice planes connected with other axes. These lattice planes, (1 0 0), (0 1 0), (0 0 1), etc., are overlapped by rotating to 90 degrees. They are called the **equivalent plane**.

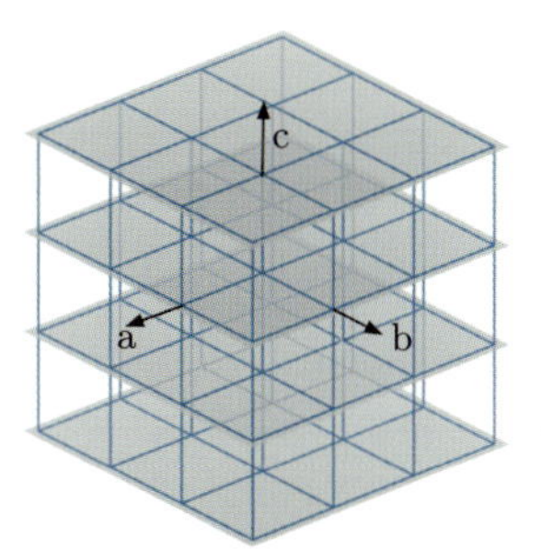

(a) (0 0 1) lattice plane of cubic

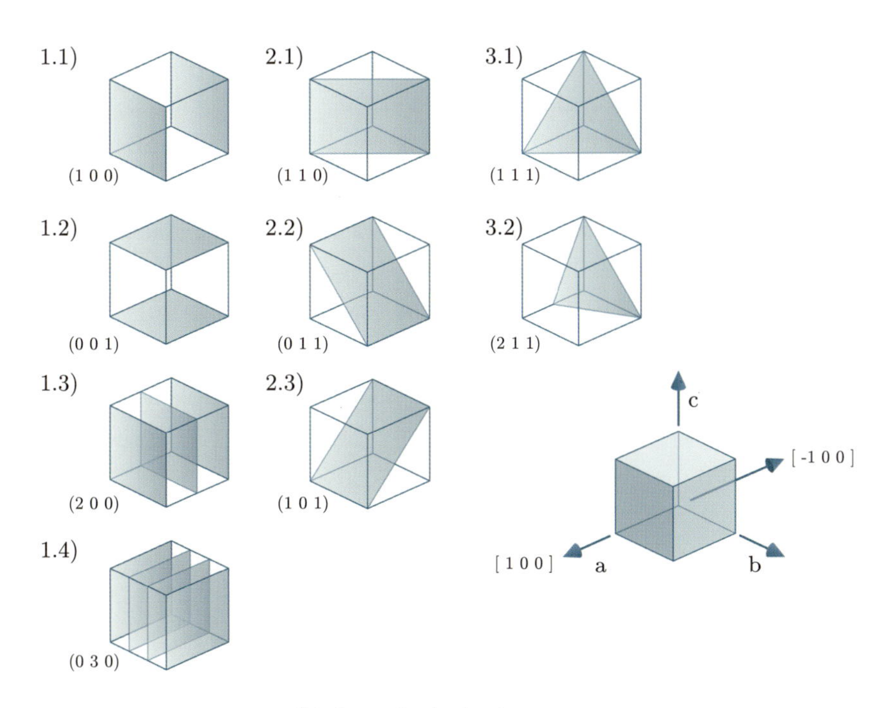

(b) Some other lattice planes

Figure 1.5 Crystal lattice and some lattice planes of cubic crystal.

Panel(1.2) of Figure 1.5 (b) : The lattice plane is (0 0 1). It is the equivalent plane with Panel(1.1).

Panel(1.3) of Figure 1.5 (b) : It is the lattice plane with the inter-planar spacing $a_0/2$. The lattice plane is designated as (2 0 0). The direction of the plane is [2 0 0], and it is the same direction with [1 0 0]. As can be inferred from Panel(1.1), the equivalent plane with (2 0 0) is (0 2 0), (0 0 2) and (0 $\bar{2}$ 0) and so on.

Panel(1.4) of Figure 1.5 (b) : It is the lattice plane with the inter-planar spacing $a_0/3$ along the $\boldsymbol{b}$-axis. We call the lattice plane as (0 3 0). In addition to this, there are the lattice planes (0 n 0) with the inter-planar spacing a_0/n ($n = 1, 2, 3 \ldots$.), although they are not shown in the figure. The direction of these lattice planes are [0 1 0].

Panel(2.1) of Figure 1.5 (b) : We consider a set of parallel lattice planes which are intersected at the position a_0 of the $\boldsymbol{a}$- and $\boldsymbol{b}$-axis, as shown in the figure. The inter-planar spacing is $a_0/\sqrt{2}$. The lattice plane is (1 1 0), and the direction is [1 1 0]. In addition, the lattice plane with a half and 1/3 of the inter-planar spacing of (1 1 0) is expressed by (2 2 0) and (3 3 0), respectively.

Panel(2.2) and (2.3) of Figure 1.5 (b) : As the equivalent plane of (1 1 0), there are (0 1 1) and (1 0 1). With the same method, the lattice planes belonging to the equivalent plane of (2 2 0) are (0 2 2) and (2 0 2), and so on.

Panel(3.1) of Figure 1.5 (b) : The lattice plane passes at the lattice point of a_0 of $\boldsymbol{a}$- and $\boldsymbol{b}$- and $\boldsymbol{c}$-axis. The inter-planar spacing is $a_0/\sqrt{3}$. This lattice plane is (1 1 1). The normal vector is the direction of [1 1 1]. There are ($\bar{1}$ 1 1), (1 $\bar{1}$ 1), and ($\bar{1}$ $\bar{1}$ $\bar{1}$), *etc*, as the equivalent plane. The lattice plane with a half inter-planar spacing of (1 1 1) is (2 2 2).

Panel(3.2) of Figure 1.5 (b) : The plane intersects $a_0/2$ at $\boldsymbol{a}$-axis and a_0 at $\boldsymbol{b}$- and $\boldsymbol{c}$-axis. The lattice plane indicates (2 1 1). The inter-planar spacing is $a_0/\sqrt{6}$. The normal vector of this lattice plane is [2 1 1], and the equivalent plane $\{2\ 1\ 1\}$ to the $(2\ 1\ 1)$ plane has 24 planes including back-side planes.

We can guess a relationship between the inter-planar spacing and the lattice plane $(h\ k\ l)$ based on the rule of the Miller indices, where, h, k, l are integers. $(h\ k\ l)$ plane intersects the position of $1/h$ at $\boldsymbol{a}$-axis, $1/k$ at $\boldsymbol{b}$-axis, and $1/l$ at $\boldsymbol{c}$-axis. Then, the inter-planar spacing (d_{hkl}) is given by Equation (1.1).

$$d_{hkl} = \frac{a_0}{\sqrt{(h^2 + k^2 + l^2)}} \tag{1.1}$$

From the Equation (1.1), we can assign the lattice plane $(h\ k\ l)$ corresponding to each Bragg reflection in diffraction pattern of Figure 1.2. The details are explained in next section.

1.4 Diffraction phenomena of X-rays by a crystal

1.4.1 Bragg reflection

We have shown the experimental results of CsCl, α-Fe, and Al in Figure 1.2 (a), (b), and (c), and mentioned that the numbers of Bragg reflections are observed depending on the sample materials. In this section, we start with reviewing the Bragg's law and examining the interplanar spacing d_{hkl} by submitting the diffraction angles observed in the Bragg's equation. Then, as the next step, let's confirm whether the results are in agreement or not with the calculations obtained by using the known data of lattice constant.

The wavelength of X-rays λ is usually a fixed value. In the present case, the wave length is $\lambda = 1.542\,\text{Å}$, since Cu X-ray tube has been used. Let us further consider the Bragg reflection happened by $(h\ k\ l)$ lattice plane of which inter-planar spacing is given by d_{hkl}.

$$2d_{hkl} \sin\theta = \lambda \tag{1.2}$$

where θ is the angle between the incident X-rays and $(h\ k\ l)$ lattice plane. Equation (1.2) is easily rewritten as,

$$\sin\theta/\lambda = 1/2d_{hkl} \tag{1.3}$$

From the Equation (1.3), we can see that $(\sin\theta/\lambda)$ is inversely proportional to d_{hkl}. This fact suggests that the observation of the Bragg reflections with high scattering angle provide valuable information about lattice plane having narrow inter-planar d-spacing in crystal. Equation (1.3) signifies that the inverse value of the inter-planar spacing $1/d_{hkl}$ is directly given by $\sin\theta/\lambda$ obtained by normalizing $\sin\theta$ by the wavelength λ used, although the values θ or $\sin\theta$ are experimentally obtained.

By the way, let the inter-planer spacing (d-spacing) d_{hkl} given by Equation (1.1) substitute into Equation (1.3). We have:

$$\sin\theta_{hkl}/\lambda = (1/2a_0)\sqrt{(h^2 + k^2 + l^2)} \tag{1.4}$$

where, θ_{hkl} has been used instead of θ in order to be clear about the Bragg angle which corresponds to the particular hkl-lattice plane. This equation indicates also that the Bragg angle observed increases with the increase of $\sqrt{(h^2 + k^2 + k^2)}$. In addition, indexing to the observed reflections can be easily found by taking square of both the sides of above equation. On the contrary, if the hkl indexing is correct, the lattice constant a_0 of unit cell in the crystal can be seen to be obtained from any of the Bragg reflections.

One of the techniques of hkl-indexing is to confine first our attention to the lattice plane of the biggest d-spacing. As for example, if taking up the first reflection of CsCl and indexing it as a $(1\ 0\ 0)$ plane, we obtain immediately the inverse of the lattice parameter $(1/a_0)$. If a subsequent reflection can be applied by indexing gradually a little higher hkl, simultaneously, we can compare all the values of $(1/a_0)$ obtained from each Bragg reflection. If the same lattice parameter a_0 is obtained from all the Bragg reflections observed, it can be said that the hkl indexing is correct. By applying this analysis, several numerical results obtained for CsCl are listed in the Table 1.1 (a). The raw a(obs) shows the lattice constant a_0 obtained from respective Bragg reflection. The average value is 4.125 ± 0.006 Å, which is in very good agreement with a(calc) = 4.123 Å which can be obtained from the literature within the experimental error. In the analysis, the most difficult point is to assign a correct index to all the Bragg reflections observed.

Table 1.1. Indexing of Bragg reflections and calculated lattice parameter obtained from observed value $\sin\theta/\lambda$

(a) CsCl (a(calc)=4.123Å) $\lambda = 1.540593$Å
Referred to the database of Crystallographic Society of Japan

hkl	100	110	111	200	210	211	220	300	221	310	311
2θ (obs)	21.494	30.616	37.717	43.844	49.341	54.445	63.785	68.173	68.173	72.409	76.555
$1/2d_{hkl}$	0.12	0.17	0.21	0.24	0.27	0.30	0.34	0.36	0.36	0.38	0.40
d_{hkl}(obs)	4.131	2.918	2.383	2.063	1.845	1.684	1.458	1.374	1.374	1.304	1.243
a(obs)	4.131	4.126	4.128	4.126	4.127	4.125	4.124	4.123	4.123	4.124	4.124

hkl	222	320	321	400	410	322	411	330	331	420	421
2θ (obs)	80.643	84.646	88.661	96.709	100.728	100.728	104.803	104.803	109.007	113.36	117.69
$1/2d_{hkl}$	0.42	0.44	0.45	0.49	0.50	0.50	0.51	0.51	0.53	0.54	0.56
d_{hkl}(obs)	1.190	1.144	1.102	1.031	1.000	1.000	0.972	0.972	0.946	0.922	0.900
a(obs)	4.124	4.125	4.125	4.123	4.124	4.124	4.125	4.125	4.124	4.123	4.125

(b) α−Fe (a(calc)=2.872Å) $\lambda = 1.540593$Å
Referred to the database of Crystallographic Society of Japan

hkl	100	110	111	200	210	211	220	300	221	310	311
2θ (obs)	–	44.565	–	64.94	–	82.17	98.9	–	–	116.69	–
$1/2d_{hkl}$	–	0.25	–	0.35	–	0.43	0.49	–	–	0.55	–
d_{hkl}(obs)	–	2.0315	–	1.4349	–	1.1721	1.0138	–	–	0.9049	–
a(obs)	–	2.873	–	2.870	–	2.871	2.867	–	–	2.862	–

(c) Al (a(calc)=4.052Å) $\lambda = 1.540593$Å
Referred to the database of Crystallographic Society of Japan

hkl	100	110	111	200	210	211	220	300	221	310	311
2θ (obs)	–	–	38.442	44.6732	–	–	65.077	–	–	–	78.2
$1/2d_{hkl}$	–	–	0.21	0.25	–	–	0.35	–	–	–	0.41
d_{hkl}(obs)	–	–	2.340	2.027	–	–	1.432	–	–	–	1.221
a(obs)	–	–	4.053	4.054	–	–	4.051	–	–	–	4.051

hkl	222	320	321	400	410	322	411	330	331	420	421
2θ (obs)	82.414	–	–	99.069	–	–	–	–	111.996	116.553	–
$1/2d_{hkl}$	0.43	–	–	0.49	–	–	–	–	0.54	0.55	–
d_{hkl}(obs)	1.169	–	–	1.012	–	–	–	–	0.929	0.906	–
a(obs)	4.050	–	–	4.050	–	–	–	–	4.050	4.050	–

If a wrong index is assigned, however, it is not difficult to judge because there happens a discrepancy with the a(obs) which can be given from another Bragg reflection. Table 1.1 (b) and (c) show the result for α-Fe and Al, respectively. You will see that the Bragg reflections of CsCl appear to all the lattice planes, but those of α-Fe and Al are not simple. We must understand why such a phenomenon happens. If you don't understand its reason, you will encounter a difficulty in assigning indexes to all the Bragg reflections observed. This subject will be taken up in the next section.

1.4.2 Extinction rule

In order to understand this section, it is assumed for reader to be familiar about the phase of a wave and also the intensity of the wave that is proportional to the square of the amplitude of the wave. The most basic idea about X-ray diffraction has been summarized in Appendix A, including above topics. Thus, it has been assumed to refer to it.

The Figure 1.6 (b), (c) and (d) show the projection views to the (0 1 1) plane along the [0 1 1] direction for several unit cells of CsCl, Fe, and Al, respectively. The Figure 1.6 (a) describes that incident X-ray wave and scattered X-ray wave are in a condition that the Bragg reflection occurs by the (1 0 0) plane of a spacing $d_{100} = a_0$. The path difference between the X-ray wave scattered from the top lattice plane and that from the second one is $2d_{100} \sin \theta$. When the path difference is equal to just one wavelength λ of the X-ray wave, the two scattering waves, one from the top lattice plane and the other from the second lattice plane, overlap completely, as the result of interference of each other. This is the condition in which Bragg reflection occurs by the (1 0 0) plane. In all the three crystals, we are considering, however, an additional atomic layer exists in a half of the (1 0 0) plane as seen in the Figure 1.6 (b), (c) and (d). In α-Fe and Al crystal except CuCl, the same kind of atoms occupy on such a lattice plane by the same amount. In Figure 1.6 (a), X-ray wave scattered from

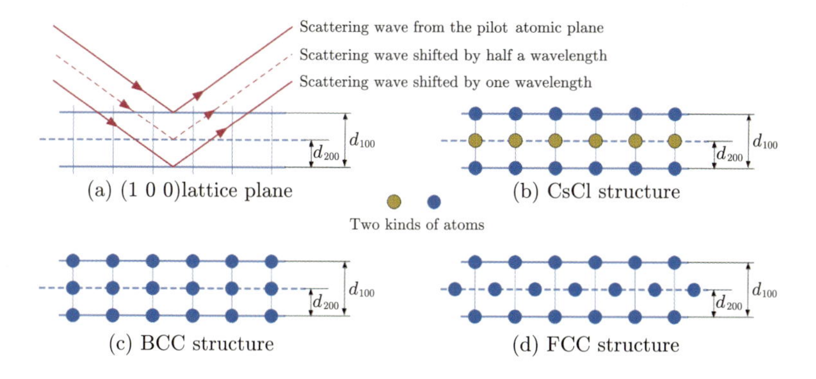

Figure 1.6 Projection view of cubic crystal along the [1 1 0] direction.

this plane is shown by a dotted line. The X-ray wave scattered along the direction of the Bragg reflection from upper lattice plane and that from the middle lattice plane have a path difference of the half wave-length; the two waves, the solid line and the dashed line in the Figure 1.6 (a), are canceled out as the result of interference of each other. Thus, the Bragg reflection from the (1 0 0) plane will not occur.

In the CsCl ionic crystal, the upper lattice plane is occupied by Cs ions while the middle lattice plane by Cl ions. The X-ray wave scattered from the Cs ion's layer has different amplitude from that scattered from Cl ion's layer. Those two waves scattered along the Bragg reflection are of also the different phases but of the different amplitudes. Thus, they are not completely canceled out. But a wave is remained of which amplitude is given to be the difference between the amplitudes of the two waves. As the intensity of a given wave, in general, is proportional to the square of its amplitude, it will be seen that the Bragg reflection from the (1 0 0) plane is weak and not disappeared for CsCl crystal but disappeared at all for all the BCC and FCC crystals, in agreement with the Figure 1.2 (a) (b) and (c).

As other example, consider the (1 1 0) plane. The CsCl is constituted by the lattice plane which is alternately occupied by Cs and Cl ion. There is no atom on the half plane of the lattice plane. Thus, the Bragg reflection appears really, because of no cancellation. This situation is the same for the BCC crystals. A strong Bragg reflection is appeared really for the BCC crystal. On the contrary, there exists the same type of atoms with the same atomic density (the number of atoms per unit area) on the half plane of the (1 1 0) lattice plane for the FCC crystals. These facts are reflected in the Table 1.1 (c): that is no diffraction peak from (1 1 0) plane is observed.

Therefore, it can be said that the Bragg reflection does not surely occur for all of the lattice planes and there are reflections of which intensity becomes weak and also becomes zero. This is true even for cubic crystal. This is called **the extinction effect**. You would be able to understand such an intensity difference existing among the Bragg reflections due to the fact that the scattering amplitude depends very much on the crystal structure (atomic arrangement in the unit cell). There exists a rule that indicates which lattice plane will be of the forbidden reflection for a given crystal. You may see from above examples that the **extinction rule** is related strongly to the structure of crystal.

Once you understand the extinction rule, you will see that the existence of forbidden reflections is a natural consequence for several lattice planes of α-Fe and Al, although such a reflection does not exist for all the lattice planes of CsCl. The next subject is to know what kind of relationship exists between the crystal structure and the extinction rule. Its answer will be given by understanding of the elementary theory of X-ray scattering/diffraction from crystal. The rule is easily obtained by taking Fourier transform of atom arrangement in unit cell. The quantity is called the **crystal structure factor**[6).

1.5 X-ray absorption by sample crystallites

Comparing the intensities of three diffraction patterns seen in the Figure 1.2 (a), (b) and (c), the intensity of (a) is around 7000 cps, that of (b) is $\sim$1000 cps and that of (c) is $\sim$15000 cps. A question comes out why such intensity difference has happen among the three samples we have measured. Although this is not so serious subject, let' keep understanding a bit about it. In order to make it clear you need to understand the subject of this section.

The degree of X-ray absorption by material is different depending on the wavelength of X-rays used (you may use the "energy" instead of wavelength). The range of wavelength of the X-rays which is commonly used in diffraction experiment is a bit longer than that of X-rays which is used in medical radiography. Therefore, X-ray is often considered to be very transparent, but in the normal material can not ignore the absorption.

When the X-rays with a certain wavelength are incident to a material, the X-rays can not penetrate deeply into the material, because the X-rays are absorbed more or less by it. If the depth by which X-rays can penetrate into the material increases, the volume of the material contributing to the X-ray scattering will also increase. In other word, for the material which has a large absorption, the volume of material contributing to the X-ray scattering/diffraction will be small, leading to the result that strong X-ray intensity is not expected to be observed. The absorption factor, therefore, is an important indicator to dominate the scattering intensity.

Generally, when the X-rays with intensity of I_0 propagate the distance t through a material, the intensity I is attenuated by a factor $\exp\{-\mu t\}$ so that $I = I_0 \exp\{-\mu t\}$, where μ is the linear absorption coefficient of the material. This parameter μ depends on the wavelength of the X-rays used

Table 1.2 Calculated value of μ and $1/\mu$ for CsCl, α-Fe, and Al crystal.

	μ (cm^{-1})	$1/\mu$ (cm)
CsCl	428	0.0023
α –Fe	964	0.0010
Al	49.0	0.020

μ : the linear absorption coefficient

and the kind of materials. In order to judge large or small absorption for a given material it is often classified the material by using its penetration depth t for a given X-rays in such a way that the intensity is decreased by a factor $1/e$. It corresponds to $\mu t = 1$ so that such a thickness $t = 1/\mu$. The linear absorption coefficients for the three materials are listed in Table 1.2 for the X-rays with the wavelength of 1.542 Å. As seen from the Table the ratio of $1/\mu$ is 1 : 0.4 : 8.7 for CsCl : α-Fe : Al. As a result, we see that the diffraction intensity of Al is significantly large compared with other materials.

The X-ray absorption coefficient does not immediately need in analyzing a given material, but it suggests us how the important to know about the physical property of the X-rays used and also the material which we deal with. In this book, the more detailed description about absorption is given in the Section 5.5, "Setting up of basic experimental tools" of the part 2; please refer to it.

1.6 Crystalline state in sample

The powder sample is scattered flatly as possible on the sample stage of an X-ray diffractometer, and the measurement is made under the condition that the incident X-rays and diffracted X-rays are always symmetric with respect to the sample surface, as shown in the Figure 1.1. There are a lot of crystallites in a region irradiated by X-rays. Among those crystallites, only crystallites of the *hkl* plane which is parallel to the sample surface contribute to the Bragg reflection.

Therefore, a fact that the Bragg reflection of *hkl* is observed at a certain incident angle can be paraphrased as follows: crystallites of which *hkl* lattice plane is parallel to the sample surface exist sufficient amount in the sample. As seen in the diffraction pattern of the Figure 1.2, all the Bragg reflections are indeed observed from three samples. This fact indicates that the orientations of the crystallites presented in those samples are evenly distributed. Preparation of powder sample mentioned above is a technology that has so far been used in order to get a good experimental result.

Among the three samples we have measured, a sketch describing powder state of α-Fe and Al samples is shown in the Figure 1.7 (a), (b), and (c). The Figure 1.7 (a) shows an enlargement of powder sample on the sample stage so as to be able to see aggregation state of grains and Figure 1.7 (b) shows an enlargement of one grain. As shown by this figure, one grain is usually formed by several tiny crystallites each of which is of different size and different orientation. One piece of crystallites which extracted from **one grain** is shown in Figure 1.7 (c), its size and orientation are arbitrary. Hearing the word "crystal", we are in a tendency to imagine a large single crystal. Crystalline materials we can have in hand, however, are quite different from above two states (powder state and single crystalline state). They are in a variety of aggregated states of crystallites which are of quite

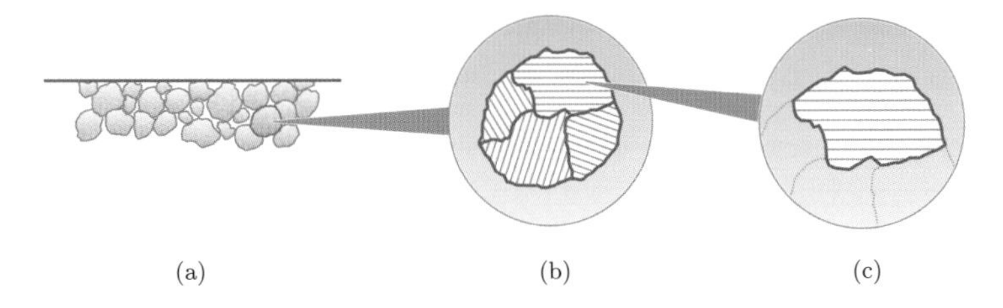

Figure 1.7 Microscopic structure of sample

different sizes and specific orientations, as explained by using metal samples in above.

Thus, we have so far paid a special attention to the preparation of sample for the measurement of the X-ray diffraction. The powder sample has been especially utilized as the size distribution of crystallites to be uniform as much as possible. If such an ordinary material was used as our sample without making it in powder state, what sort of diffraction pattern could be obtained? It may happen that the diffraction line from a certain lattice plane could be strong whereas that from other lattice planes could be weak or zero. This is probably due to the fact that the size and the orientation of crystallites are indescribably random. Such a diffraction pattern may not be useful for identification of the material, but may be utilized as a device to study the aggregation state of crystallites in the material.

1.7 Crystalline texture

Most of materials are formed by several very tiny crystallites, which have such a characteristic property that their sizes are less than micrometer, although a glass material is included also. Materials formed by such a texture are called **crystalline texture**.

Material is made of a lot of crystallites. There is a material that is made from a single type of crystallite, but there are also those that are made from several different kinds of crystallites. The crystallites have their own shapes so that there is a case that the most of crystallites are oriented along the some particular direction of crystallite. But also there is another case that the crystallites have a characteristic size distribution. The difference of the size distribution of crystallites will be considered to influence significantly their material properties. It would be advisable to know the size distribution, however, which can be obtained by using the X-ray diffraction method. As presented in the previous section, the width of the Bragg reflection is closely related to the average size of crystallites in material. The detailed analysis of the profile of some particular Bragg reflection has been the subject to study. By analyzing the profile, the characterization as well as the clarification of crystalline texture has been fairly well progressed.

1.8 Qualitative and quantitative analysis

Researchers and the engineers, who are skillful in the analysis of the diffraction patterns, can easily presume a crystal structure which is appropriate to the observed diffraction pattern. Because these peoples estimate final answer by comparing the data obtained to own database which is stored around them. In order to confirm a presumed crystal structure, usually the diffraction data should be compared with database. In this process, the most important point is in the degree of agreement of the diffraction angles with database, rather than in the comparison of the intensity ratio. Consistency of the intensity ratio with the database can not be said so important. Its reason will be given in Chapter 3 and Chapter 4. Therefore, we sometimes assign the intensity observed with 5-point scale; that is very strong, strong, medium, weak and very weak, which are abbreviated as VS, S, M, W, VW. The data analysis by such a method is the **qualitative analysis**. According to the need, the lattice parameter which is calculated by the diffraction angle obtained can be compared with the value reported in database. Any database is available to use, so you may utilize even your own database.

In addition to the qualitative analysis, it may happen necessity to confirm quantitative value of inclusions in some substance. In such a case, X-ray diffraction is a useful also for the quantitative analysis to apply. It will happen to encounter an unknown substance that cannot be identified even by using available database. It will be requested to find out to which crystal system the substance belong based on the diffraction pattern obtained. A more detailed analysis will be furthermore requested until atomic arrangement in the unit cell can be obtained. This is called **crystal structure analysis**. Besides, the evaluation of **crystalline texture**, mentioned in the previous section, can be performed by X-ray diffraction.

At the present time, the database accumulated in several institutions can be used by anybody. Convenient software is also provided so that you may use easily one of those databases by installing it in your computer. After you have learned the techniques how to measure and how to analyze the data obtained, you are in a situation that any material of your choise can be analyzed, because convenient software system which has been almost well prepared can be used. However, the author hopes that you extend your study by developing more advanced technology which exceeds the limits of such current systems. Under a special environment condition, a variety of new crystalline materials, of which crystal structure have never been investigated, become available, and their application studies are also becoming popular. The analysis of the crystal structure and the evaluation for those materials will be requested also by overcoming the limitation of the presently available system in the X-ray diffraction. Even in such a case author is sure that you will get proper answer if there is in attitude that you think to go back and you add modifications on your device and software on the basis of the principle of the X-ray diffraction and the X-ray optics. The subject of this book is to clarify the technology of X-ray diffraction using MiniFlex 300/600 by showing practically the qualitative analysis, but the evaluation of crystalline texture is included to some extent.

Annotations

1) About the words **scattering**, **diffraction**, and **reflection**: In describing the principle of measurement by X-ray diffractometer, as shown in the Figure 1.1, the following expressions are used: when X-ray beam is irradiated to a material, the X-rays are strongly reflected from the irradiated region along several specific directions. In ad-

dition, the words "scattering angle 2θ" and "scattering plane" have been used in the various sections. A word X-ray diffractometer is used also for the device to measure the scattering X-rays. Everyone will wonder the reason to say diffractometer without saying X-ray scattering measurement device. The word diffraction line or peak is often used instead of Bragg reflection in the subsequent chapters. As these words have been used routinely, it will be explained how to use three words, scattering, diffraction and reflection by showing some examples.

The X-rays are the electromagnetic wave with the wavelength which is very much shorter than that of visible light. The X-rays behave as waves as well as particles which are referred to photons. Here, the scattering and diffraction phnomena of the electromagnetic wave is discussed. We consider a scattering phenomenon of X-rays from an object of which size is very small compaired with the wavelength such as an electron which is originaly rest. The electromagnetic plane wave with the wavelength of λ collides with the electron. As the electron oscillates by the alternating electric field of the electromagnetic plane wave, a dipole oscillation with the same frequency as that of the incident electromagnetic wave is produced. Such a dipole oscillation emits the electromagnetic spherical wave with the same wavelength from the point where the electron was rest. This is the X-ray scattering by an electron and is particularly called the **Thomson scattering**. This means that scattering wave has the same wavelength with that of the incident wave. In other words, it means that both the electromagnetic waves before and after the scattering have the same energy so that no energy transfer has happened during this scattering process.

We say this is the elastic scattering. Of course there is inelastic scattering phenomenon in the X-ray scattering by an electron. That is a case that electron is flicked away by the X-rays. In understanding this phenomenon, it is convenient to think as a collision of a photon by an electron. The X-ray photon loses the energy which is the same amount of a kinetic energy which the electron got in the process of the scattering. As a result, the scattered X-rays become of a long wavelength. In other words, energy transfer from the photon to the electron happened during the colliding. This is a tipical example of **inelastic scattering** and is called the **Compton scattering**. You will understand that in describing all above phenomena, elastic and inelastic scattering, the word of scattering is appropriate to use.

As the second subject, let us consider the X-ray scattering from an atom which is regarded as an aggregation state of electrons. In above section, it has been shown that there are two scattering processes: one is the elastic scattering in which there is no exchange of energy between X-ray waves and an electron, and another is the inelastic scattering process accompanying energy exchange. In the scattering of X-rays by an atom, there can occur also a kind of inelastic scattering. Since a slight change of the configuration of electrons around nucleus can happen by X-rays. From now on, however, we take up the elastic scattering process only.

Thus, we treat X-rays as to make the Thomson-scattering with the electrons which are distributed almost spherically symmetric around nucleus of atom. The scattering from more than one electron can be treated as the summation of the scattering waves from each scatters under a condition that all the scattering waves are coherent to one another. The X-rays we are concerned with satisfy this condition. In such a case, the point we have to take care is that all the scattering waves from each electron have different phases due to the different location of electrons which are the scatters. As a result, after taking care of all of them the scattering from an atom is found to be forward scattering. The scattering to the backward direction from the atom is small compared with forward direction. In order to describe this phenomenon, the word of diffraction is not used. Instead, the word of scattering is in common. There is a word "**atomic scattering factor**" as an example.

Here, let us consider the scattering phenomenon by two atoms, possible to regard as two scattering bodies. The distance between the two atoms is assumed to be the same order of magnitude with or less than the wavelength of the incident X-ray waves so that two waves scattered from two atoms are able to interfere each other. As the result of interference of two scattering waves, it happens that X-rays are fairly strongly scattered in a periphery of a certain scattering angle but not so much in other angles. This phenomenon we have had now is called X-ray diffraction by an atomic pair. The angle representing a specific direction in which X-rays are scattered strongly is also called **diffraction angle**, instead we use the word "scattering angle" for other angles in which X-rays are not scattered so much.

The meaning of "reflection" in the word of "Bragg reflection" is slightly different from a normal one. We consider the scattering phenomenon of X-rays with a definite wavelength from several atomic planes which are piled up. The scattering waves from the respective atomic planes overlap one another. As the result, one synthetic wave is produced, of which amplitude is modulated depending on the direction to go out. If the incident waves come in from such a direction as to satisfy the Bragg condition (among the parameters: wavelength λ, inter-atomic spacing d and the incident angle θ), the amplitude becomes large in the direction which is just symmetric with respect to the direction from which the incident wave comes in. At first sight, this phenomenon is understood as that a part of incident X-rays are reflected from the interface of medium and the rest of X-rays are refracted at that surface and then go straight. So this phenomenon is very resemble with a reflection phenomenon of visible light. Adopting particularly the name of a famous discoverer, it has been called the Bragg reflection. This is a very meaningful word that is represents the physical phenomena well.

2) Besides the $\theta - 2\theta$ measurement method of the X-rays scattered under a condition of a symmetrical reflection with respect to the sample surface, there is $\theta - \theta$ method. In this case the sample stage is kept as horizontal. Both the arm of the incident X-ray beam and the arm holding the detector rotate around the center on the sample surface

but to the opposite directions each other.

3) In visible light reflection, this angle is called the glancing angle, and the incident angle is its complementary angle. In X-ray diffraction, the angle is called the incident angle.

4) D/teX Ultra is a 1D silicon strip detector. Using the one Dimensional detector (1D), you can measure precisely as well as effectively the scattering X-rays. In MiniFlex 300/600 with this detector, the angle range able to measure in 2θ is from 2 degree to 140 degree. If adjustment is made, it is possible from 1 degree. The step width in the measurement is $\Delta 2\theta = 0.02$ degree, and its speed is 0.6 degree/sec. The details are described in X-ray Optics of the Chapter 7.

5) The monochromater of HOPG will be described in Monochromatization of X-rays of Chapter 7. It is sufficient in this Chapter, if you could understand it as a simple device which can select the X-rays with a specific wavelength.

6) Fourier transform is the words used in mathematics which came out suddenly. It is necessary for you to understand it if you would like to be familiar with the diffraction crystallography. But, a beginner may want to know how much mathematical knowledge is required. Roughly speaking, the necessary mathematics is vector operation, simple wave equation and Fourier transform. As some necessary comment will be tried at each time, it is recommended that you learn the necessary mathematics along with reading this book.

Chapter 2 Integrated intensity and crystal structure

In the previous chapter, it was shown that X-ray diffraction pattern which comprises a limitted number of Bragg reflections is observed from a given sample with the use of a X-ray diffractometer. It has also shown that the sample can be identified what kind of material is by comparing the Bragg angles and its relative intensities observed with the existing database. This is usual process in the qualitative analysis of a given sample by X-ray diffraction. In this chapter, we show a basic expression to reproduce diffraction pattern observed, and on the basis of it we discuss about the relation between the intensity and the crystal structure of the sample. In addition, the diffraction pattern is shown to be reproduced by numerical calculation based on the expression, if the atomic configuration in the crystal is given.

2.1 Powder X-ray diffraction pattern

Polycrystalline powder sample[1)] is considered to be composed of aggregates of many crystallites which are oriented very randomly. If irradiated a parallel X-ray beam to such a sample and set an X-ray film or a 2-dimensional detector so as to be perpendicular to the direct beam at the position behind the sample, as shown in Figure 2.1 (a), the X-rays scattered forward direction from the sample can be recorded by the detector. Diffraction pattern observed is consisted of many concentric circles. It is called the **powder diffraction rings**.

When such an experimental fact is obtained, it means conversely that there are many crystallites oriented randomly in the sample. In other

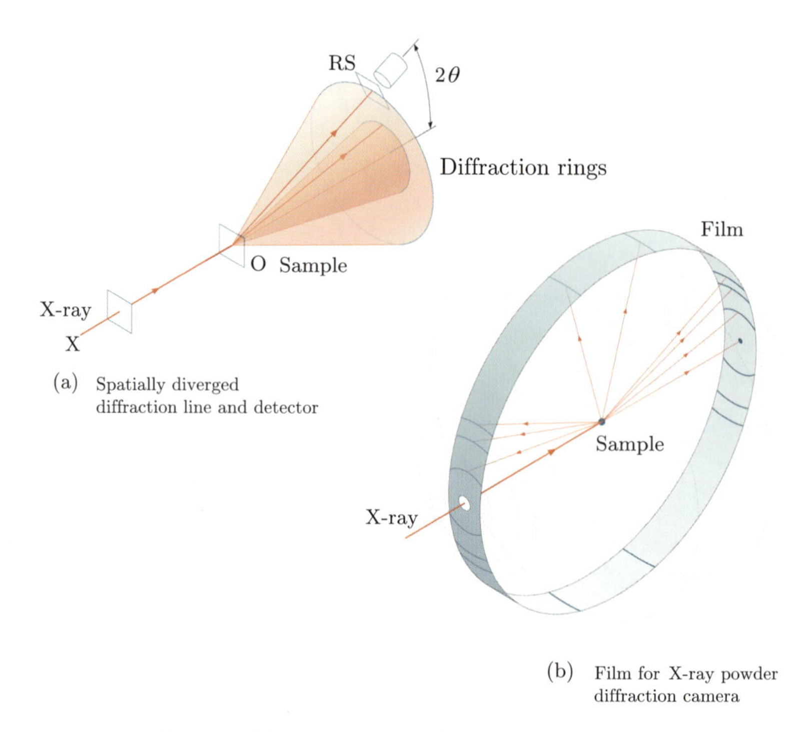

Figure 2.1 Observation principle of X-ray powder diffraction pattern.

words, it means that all the lattice planes which can be indexed by hkl exist surely around the direct beam and also satisfy the Bragg condition. The X-rays diffracted from those crystallites will form a cone shape with the vertex angle of twice the diffraction angle $2\theta_{hkl}$.

Instead of using a plate like film, if winded cylindrically a strip-shaped film around the sample as shown in Figure 2.1 (b), all the diffraction lines of which diffraction angle 2θ are in the range from 0 degree to ± 180 degrees are observable. The diffraction instrument equipped with this type is called the **X-ray powder diffraction camera** or **Debye-Scherrer camera**. This camera is compact and easy-to-use, but has not been currently used. Because X-rays film has not been produced any more from the reason that there is a waste disposal problem of developing solution for any type of films. When taking into account of the fact that the spatial resolution of the film is of the order of 1 μm^2 square and is very good compared with any available 2D detectors and that even though there is a big demerit in using it because of very low sensitivity for X-rays, it is regrettable that the film cannot be used.

Imaging plate (IP) has so far been utilized from the time it was developed about 40 years ago by Japanese film maker as an alternative to X-ray film. X-rays are recorded in a plate which is called Imaging Plate and can be read out as an electric signal by illuminating laser light. Its spatial resolution is fixed according to the size of the laser beam. At present, the size is about $25 - 50\,\mu m^2$, fairly sufficient. The imaging plate has been used for X-ray medical equipment so that some of you might know it already. A few diffraction equipments accepting imaging plate have been manufactured by RIGAKU: that is RINT and RINT-Rapid. In RINT, a cylindrical IP is utilized ,which is able to observe the diffraction patterns from a single crystal. The RINT-Rapid is an equipment for the X-ray powder diffraction. As an example, a diffraction pattern obtained by using RINT-Rapid is shown in the Figure 2.2, where the sample is Al_2O_3. All the diffraction lines are of elliptical. Its reason is due to the use of cylindrical camera. In the analysis, however, there is no difficulty at all. In this figure, a white round shape is also seen in the center that is a shade of the beam stopper. In order to make sure the center of diffraction pattern a dotted line has been drawn in the figure. We can confirm that all the diffraction rings are observed very uniformly around a center. The advantage of such a 2-dimensional detector as IP is in that we can confirm the uniformity of the distribution of crystallites in the sample.

As for the data obtained by **X-ray diffractometer** such as MiniFlex 300/600, a part of the diffraction pattern is recorded: the part is obtained using a slit with a certain width along equator line of the diffraction pattern.

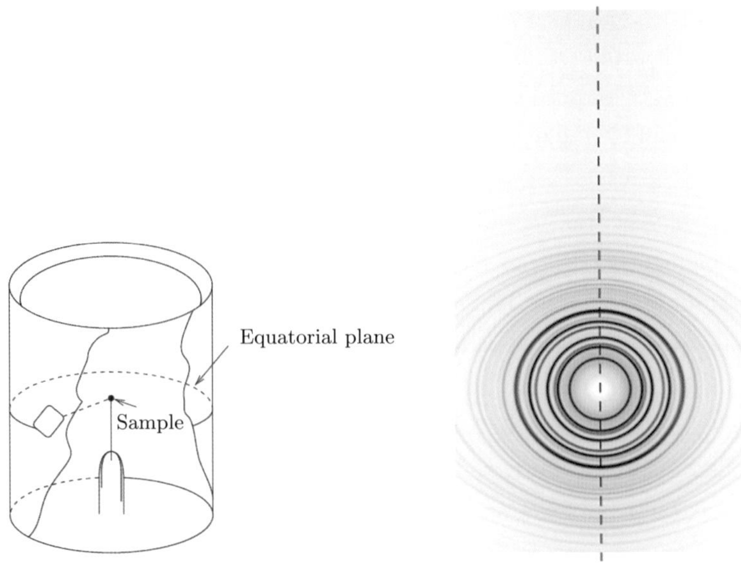

(a) Layout of cylindrical camera

(b) Diffraction pattern of Al_2O_3 (a dotted line through a center is the equator line.)

Figure 2.2 Powder X-ray diffraction pattern of Al_2O_3 observed by IP film of cylindrical type.

All the diffraction patterns shown in Chapter 1 are the results obtained by MiniFlex 300/600 which is one of diffractometers commercially available : it suggests that the diffraction pattern corresponds to the measurement by a finite slit along the dotted line of diffraction rings which you can see in the Figure 2.2 (b). If you did not care about it, you might have a difficulty to understand the meaning of the intensity equation that will be given later sections.

2.2 Scattering/diffraction from lattice plane of crystal

Figure 2.3 is a two dimensional drawing presenting symbolically atomic arrangement in a crystal. The circles represent the atoms that comprise the crystal lattice. As shown in the figure, d indicates the interplanar spacing. We examine the scattering phenomenon resulting from the irradiation of X-rays with a wavelength of λ onto the lattice plane at an angle of θ.

As the first step in this examination, consider the scattering of X-rays caused by the atomic plane A – A′, the first-layer[2)]. The X-ray beam entering at an incident angle θ relative to the surface is scattered by the atoms in the surface layer. It is reasonable to assume that the phenomenon observed is the same as the phenomenon resulting when the atomic plane A – A′ behaves as a flat mirror, being, in fact, identical phenomenon. Regardless of the incident angle θ, scattered waves reflect along the direction of the emergent angle equal to the incident angle. This phenomenon like a

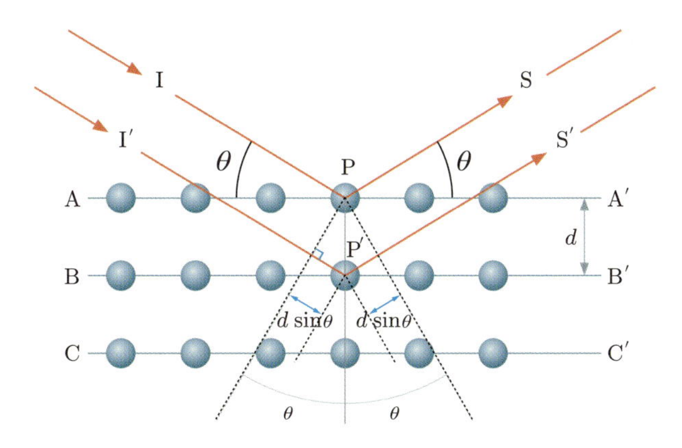

Figure 2.3 Bragg reflection by crystal.

mirror reflection occurs with any incident angle. Please take note that all incident X-rays are not reflected. A part of the X-rays are reflected from the atomic plane but another part of the X-rays pass through it.

Next, consider X-rays scattered by the atomic plane $\mathrm{B}-\mathrm{B}'$ at distance d from the aforementioned atomic plane. If the atomic plane $\mathrm{A}-\mathrm{A}'$ did not exist, atomic plane $\mathrm{B}-\mathrm{B}'$ would result in the same mirror reflection. However, when the two atomic planes cause scattering, the mirror-reflected waves from the top atomic plane and the mirror-reflected waves from the atomic plane below interfere, reinforcing each other when their phases overlap. As indicated in Figure 2.3, the path difference equals the difference between path IPS and path $\mathrm{I'P'S'}$, which can be expressed as $2d\sin\theta$. When this path difference is an integral multiple of the wavelength, the reflected waves from the two atomic planes are mutually reinforcing. The reinforcing direction θ can be obtained n as a positive integer with the following equation:

$$2d\sin\theta = n\lambda \tag{2.1}$$

This equation is based on a condition that results in the overlapping of the X-rays waves reflected from the surface atomic plane and one below it. The condition does not change even if the number of atomic planes increases. Equation (2.1) is based on a condition in which all reflected waves from each lattice plane have the same phase and reinforce one another. After this, we take the case of $n = 1$ in the Equation (2.1), corresponding to the case in which path difference between adjacent atomic layers is one wavelength. X-rays reflected in direction θ are called the **Bragg reflection**. Equation (2.1) is called the **Bragg condition**, while θ is called the **Bragg angle** [2]. This condition differs from the mirror reflection in that no reflection occurs if X-rays enter at an angle not meeting the above condition.

What is the difference between the phenomenon arising from two or three atomic planes and that arising from many atomic planes? We can not get the answer from the Equation (2.1). The difference is seen in the angular width of the Bragg reflection. It is possible to show it by calculation. The bigger the number of atomic planes, the narrower the angular width $\Delta 2\theta$ of the Bragg reflection becomes, making the condition more restrictive, as shown in Figure 2.4. It also shows the intensity of the Bragg reflection calculated against the scattering angle $\sin\theta/\lambda = 1/2d_{hkl}$. The Bragg reflection width $\Delta 2\theta$ is inversely proportional to N (the number of atomic planes). We can use this relationship to estimate the size of the crystallite ($L = Nd$) using X-ray diffraction. (This is described in further detail below.) This is an important principle to keep in mind.

A more rigorous explanation of the full width at half maximum of the Bragg angle can be obtained by focusing on the phenomenon that an atomic plane causes mirror reflection. The **Laue condition**, derived from the theory of scattering from finite crystal, which was discovered by Laue, leads to a clear explanation of the full width at half maximum.

It is interesting to see how to change the intensity profile with increasing the number of atomic layer. It is possible to calculate it if based on Laue function. The calculation is in Figure 2.4, showing the change of intensity profile with the increase of atomic layers, two layers, three layers and then 10 layers. As a result, the angle width which is satisfied with the Bragg condition depends on the number of the atomic layers N. The peak intensity of the Bragg reflection is proportionally to $(NM)^2$ [3], where M is the number of atom on one layer and N is the number of atomic layer (or lattice plane).

On the other hand, the full width at half maximum (FWHM) $\Delta 2\theta$ of Bragg reflection is inversely proportional to the number of atomic layers N. When N is large, the diffraction peaks become narrower. The relationship is one of the important principles; it is available to estimate a average crystal size $L = Nd$ using X-ray diffraction data. The peak intensity increase with square of N whereas the width becomes narrower in

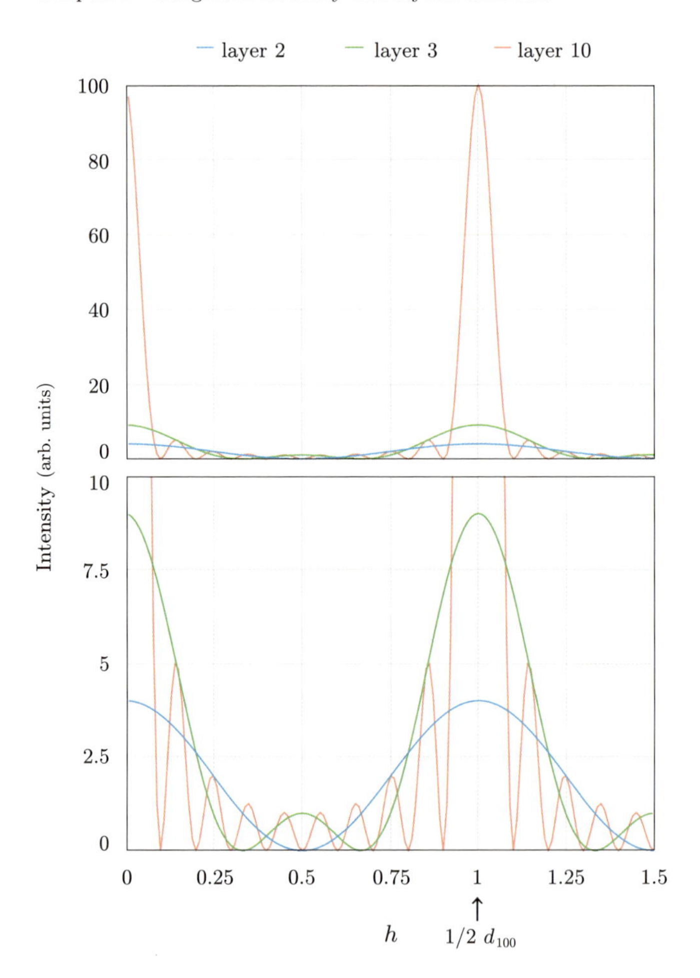

Figure 2.4 Change of (100) Bragg reflection profile with the increase of number of atomic layers. $h=2a_0 \sin\theta/\lambda$ (a_0 is the lattice parameter of cubic.)

proportion to $1/N$. It means that angle width to meet Bragg condition becomes severe.

2.3 Intensity formula for powder X-ray diffraction lines

In the previous section, we have shown a result that the peak intensity of the Bragg reflection from a crystallite of a finite size is proportional to the square of NM which represents the number of scatterers in the crystallite. In this explanation, we have not mentioned anything about the unit scatterer of which the crystallite is consisted. But the smallest unit in a crystallite is the unit cell which is consisted of number of atoms. The scattering amplitude of the unit cell is given by the **crystal structure factor** F_{hkl} which is different, depending on the index $h\ k\ l$ of lattice plane. Accordingly, the peak intensity of the Bragg reflection indexed by $h\ k\ l$ from a crystallite should be written to be proportional to the $(e^2/mc^2)^2(NM)^2|F_{hkl}|^2$. As for the Thomson scattering, $(e^2/mc^2)^2$ is a simple numerical factor so that it is omitted from the present discussion.

Now, let consider the intensity of the Bragg reflection indexed by $h\ k\ l$ which arises from a powder sample in which very many crystallites of a variety of sizes exist. As the result, we see that the intensity is also proportional to an additional factor which is dependent on the index of $h\ k\ l$. Besides, there are some other corrections due to the technique used for the measurement. Thus, it becomes necessary to take into consideration about a few correction, which will be discussed in the next paragraphs.

Multiplicity factor : Among the many crystallites in a sample, there exist always crystallites in an orientation causing Bragg reflections. There would be many those crystallites, having no relation to their sizes. Thus, we should understand that the intensity observed as a Bragg reflection is the sum of the contribution from such crystallites. It is however necessary to consider whether the number of crystallites which contribute to the Bragg reflection indexed by $h\ k\ l$ is the same as that for other Bragg reflections. This answer is "NO". The quantity representing m_{hkl} should be take into account. The m_{hkl} is called the **multiplicity** or the **multiplicity factor**.

For the sake of discussion, let assume the crystallites in our sample be of one kind and belong to a cubic system. Suppose that the h 0 0 lattice plane causes diffraction. For this Bragg reflection, 0 h 0 and 0 0 h lattice planes are also surely included in the crystallites of the orientation causing Bragg reflections. A cubic crystallite has four fold symmetry around the [0 0 1] axis. It means that the crystallite is not distinguishable, even if it rotates around the [0 0 1] axis by 90 degree, 180 degree, 270 degree or 360 degree. It means that the (h 0 0), (0 h 0), and (0 0 h) lattice planes in crystallites are equivalent. As we can also not distinguish the lattice planes by the replacement of h by -h in the above three lattice planes, we see that there are 6 equivalent {h 0 0} lattice planes in the cubic crystal. Next, consider the (h k 0) lattice plane. The (-h -k 0), (h -k 0), (-h k 0) lattice planes as well as the (k h 0), (-k h 0), (k -h 0) and (-k -h 0) lattice planes contribute to the Bragg reflection with equal probability. This also applies to the lattice planes (0 h k) and (0 -h k). In the cubic crystal system total number of the equivalent lattice planes is therefore 24, contributing to the (h k 0) Bragg reflection.

This argument is telling us that the probability by which the (h k 0) lattice plane causes Bragg reflection is 4 times higher than that of the (h 0 0) lattice plane. The number of equivalent lattice planes is usually specified in a way that $m_{h00} = 6$ and $m_{hk0} = 24$ and so on. The m_{hkl} is used as a quantity proportional to the probability to come across the (h k l) lattice plane. For all the lattice planes with $h \neq k \neq l$ in the cubic system, $m_{h00} = 6$, $m_{hhh} = 8$, $m_{hh0} = 12$, $m_{hhl} = 24$, $m_{hk0} = 24$, and $m_{hkl} = 48$ are obtained.

Table 2.1 Multiplicity factor in crystal systems.

Crystal system	Miller indices and multiplicity factor
Cubic	$\{h\ k\ l\}$:48, $\{h\ h\ l\}$:24, $\{h\ k\ 0\}$:24, $\{h\ h\ 0\}$:12, $\{h\ h\ h\}$:8, $\{h\ 0\ 0\}$:6
Tetragonal	$\{h\ k\ l\}$:16, $\{h\ h\ l\}$:8, $\{h\ 0\ l\}$:8, $\{h\ k\ 0\}$:8, $\{h\ h\ 0\}$:4, $\{h\ 0\ 0\}$:4, $\{0\ 0\ l\}$:2
Hexagonal	$\{h\ k\ l\}$:24, $\{h\ h\ l\}$:12, $\{h\ 0\ l\}$:12, $\{h\ k\ 0\}$:12, $\{h\ h\ 0\}$:6, $\{h\ 0\ 0\}$:6, $\{0\ 0\ l\}$:2
Trigonal (Rhombohedral axes)	$\{h\ k\ l\}$:12, $\{h\ h\ l\}$:6, $\{h\ h\ 0\}$:6, $\{h\ h\ h\}$:2
Orthorhombic	$\{h\ k\ l\}$:8, $\{h\ 0\ l\}$:4, $\{h\ k\ 0\}$:4, $\{0\ k\ l\}$:4, $\{h\ 0\ 0\}$:2, $\{0\ k\ 0\}$:2, $\{0\ 0\ l\}$:2
Monoclinic	$\{h\ k\ l\}$:4, $\{h\ 0\ l\}$:2, $\{h\ 0\ 0\}$:2
Triclinic	2

In the above, only a simple cubic crystal system was discussed. We should also perform calculations for tetragonal and hexagonal crystal systems. We can easily make a mistake in counting the number of equivalent lattice planes in ordinary crystal systems, if we fail to consider the symmetrical property. We must proceed carefully. Refer Table 2.2 which has been prepared by our superior scientists for all the crystal systemes.

Lorentz factor : When X-rays are irradiated onto a powder sample, X-rays are diffracted in a cone shape, with a diffraction angle of 2θ as shown in Figure 2.1 (a). Using a detector in front of which we usually set a slit of some width, we measure only the X-rays diffracted onto the equatorial plane (along the dotted line of the film shown in Figure 2.1 (b)). Thus, the amount of X-rays accepted by the detector unit is different depending on diffraction angle of 2θ. The result shows that the integrated intensity changes with $1/(\sin\theta \sin 2\theta) \equiv L$, depending on the diffraction angle 2θ. This is called the **Lorentz factor**.

Polarization factor : Since X-rays are electromagnetic transverse waves, the direction of the electrical field vector which we call polarization direction changes each time when X-rays change its propagation direction after scattering. (Such a change of the polarization is discussed in this section.)

If the X-rays collide with one electron, the electron vibrates by receiving the electric field and emits electromagnetic waves to every direction with the electron at the center. This is the scattering phenomenon of X-rays by an electron and is called the Thomson scattering. By this scattering, the polarization direction of the scattering waves changes depending on the direction we observe. Usually we treat this phenomenon by dividing the polarization direction into the vertical direction and the horizontal one. As the result, the scattering intensity from an electron is represented by $(e^2/mc^2)^2(1 + \cos^2 2\theta)/2$, where $(e^2/mc^2)^2$ and $(1 + \cos^2 2\theta)/2 \equiv P$ are called the **Thomson scattering** and the **polarization factor**, respectively. Details about the derivation of the result have been given in Appendix A.2. The result shows that the polarization factor P consists of two terms, that is the vertical term, $1/2$, which does not change before and after the scattering, and the horizontal term, $\cos^2 2\theta/2$ which changes with the scattering angle 2θ.

In the case of measurement by using MiniFlex 300/600 with an attached monochromator, the horizontal component of the polarization is again changed. If the X-rays are monochromated by the diffraction angle $2\theta_{\mathrm{M}}$, the polarization factor is given by $(1 + \cos^2 2\theta \cos^2 2\theta_{\mathrm{M}})/2$. Here, we should be cautious about the fact that $\cos^2 2\theta_{\mathrm{M}}$ is a fixed value which is smaller than 1.

Absorption factor : With the B-B method, the measurement condition requires the incident angle θ of the X-rays on a sample with a flat plate shape to be the same as the angle θ of the diffracted beams (Figure 2.3). This is called the condition of symmetrical reflection. In the following, we will look at common asymmetry and see what happens when its special condition is for **symmetrical reflection**. We can examine the absorption of X-rays when the incident X-rays with an incident angle θ_1 are diffracted by a flat-plate sample at an angle θ_2. The absorption coefficient $A(\mu)$ can be analytically obtained, although is omitted here, and is given by the following equation:

$$A(\mu) = \frac{\sin\theta_2}{(\sin\theta_1 + \sin\theta_2)\mu} \tag{2.2}$$

In the B-B diffractometer, the sample plate and detector are driven by the θ-2θ scan system to maintain the condition of symmetric reflection for a plate-shaped sample at all times. When $\theta_1 = \theta_2$, the Equation (2.2) is significantly simplified as follows:

$$A(\mu) = \frac{1}{2\mu} \tag{2.3}$$

The value remains constant, unaffected by incident angle θ. However, the value of μ can change depending on the method of sample preparation, and careful attention is required. When the value of μ is large, X-rays do not penetrate deeply, leaving only the sample surface as the valid plane. The value of $A(\mu)$ is regarded as constant and incorporated into the proportional constant. It does not affect integrated intensity in the slightest. If the value of μ is small, however, the value of $A(\mu)$ may happen to changes depending on sample thickness. Thus, the calculations must be performed correctly. If the sample surface is noticeably irregular, absorption factor is not constant throughout the measurement. It varies from the measurements in smaller scattering angles to that in large scattering angles. This is yet another factor to keep in mind.

Integrated intensity[4)] : The intensity $I(2\theta)$ of the X-rays scattered at the Bragg angle 2θ in powder diffraction pattern is proportional to the Thomson scattering $(e^2/mc^2)^2$ and also the square of the structure factor $|F_{hkl}|^2$. As mentioned before, the three correction factors, the Lorentz-polarizaton factor Lp, the multiplicity factor m_{hkl}, and the absorption factor $A(\mu)$, have to be taken into account.

The size and the number of crystallites contributing to the Bragg reflection are considered as the constant of proportionality. Similarly, the Thomson scattering factor is constant. Thus, the intensity integrated with respect to the vicinity of the Bragg peak, which is called the **integrated intensity** $J_{hkl}(2\theta)$, is given as follows.

$$J(2\theta) \backsim Lp \cdot m_{hkl} \cdot A(\mu) \cdot |F_{hkl}|^2 \cdot \varDelta\left(\sin\theta/\lambda - 1/2d_{hkl}\right) \qquad (2.4)$$

$\varDelta$-function[5)] is here introduced so as to make sure that this formula is not a continuous function of scattering-angle 2θ. This $\varDelta$-function $\varDelta(\sin\theta_{hkl}/\lambda - 1/2d_{hkl})$ is a conditional equation to indicate that Equation (2.4) is approved when the scattering angle that enables the $h\ k\ l$ lattice plane to satisfies the Bragg condition.

As seen from the Figure 1.3 all the diffraction lines are usually observed as peaks which have a width with a smooth background scattering. After removing the background, all the observed intensities forming a peak are integrated. This is the integrated intensity of the diffraction line and is used in the structure analysis of the material of the sample. It should be noted that observed peak values are not used in the analysis. Its reason is in a fact that peak value is often different at every time of the measurements. In general, not only its peak value but also its width of Bragg reflection are physically meaningful quantities, as mentioned before. Because it happens quite often that all the profiles of the Bragg reflections are observed not to be of the same shape. However, if you can confirm that all the widths of the reflections are the same, it must be quite safe to use the peak value for your analysis.

2.4 Crystal structure factor

By showing three powder X-ray diffraction patterns obtained from CsCl, α-Fe, and Al, it has been shown previously that the diffraction lines or Bragg reflections in the patterns are of the characteristic which are responsible for the atomic arrangement in the unit cell of crystallites. Thus, the responsible quantity in the intensity of the diffraction line is the scattering amplitude from unit cell. We call it **crystal structure factor** or **structure factor** in abbreviation. The formula of structure factor is given as:

$$F_{h\ k\ l} = \sum_{j=1}^{m} f_j\,(\sin\theta/\lambda)\cdot\exp\{2\pi i(hx_j + ky_j + lz_j)\} \tag{2.5}$$

In this formula, $f_j(\sin\theta/\lambda)$ is the **atomic scattering factor** of the j-th atom. It represents the amplitude of X-rays scattered in the direction of scattering angle of 2θ, when X-ray plane wave of the wavelength of λ hits the j-th atom. The numerical values of atomic scattering factor are tabulated for all kind of atoms in the International Tables for crystallography as a function of sinθ/λ, so that anybody can use it. Atomic scattering factor is a function of the scattering angle of 2θ, but 2θ changes with the wavelength of X-rays used so that it is not convenient variable to use. Atomic scattering factor is, therefore, given as a function of $\sin\theta/\lambda$ which is the variable normalized by the wavelength. The derivation of atomic scattering factor and its physical meaning are presented in Appendix A.

In the Equation (2.5), the factor $\exp\{2\pi i(hx_j + ky_j + lz_j)\}$ represents the phase difference which the X-rays scattered from the j-th atom located at the position x_j, y_j, z_j in unit cell have, compared with the X-rays scattered from an atom located at the origin. So this is called the **phase factor**.

By summing up of the scattering amplitudes from all the atoms, $j = 1, , \ldots, m$, in unit cell taking account of the phase factors, we can calculate the amplitude of X-rays scattered in the direction specified by $(h\ k\ l)$. In the manner, and the extinction rule can be understood. If such information about the structure is provided that of which atom is located at which position, you can calculate **crystal structure factor** by substituting those information into Equation (2.5). You can confirm what sort of extinction rule exists by looking at the results obtained by the calculation.

You may deeply understand the structure factor with a couple of concrete examples. Let's calculate the structural factor of CsCl and α-Fe, and confirm their extinction rules. The CsCl crystal is of a cubic structure that one CsCl molecule is in the unit cell, giving by the following coordinates of Cs and Cl atom :

Cs : (0, 0, 0)   Cl: (1/2, 1/2, 1/2)

Substituting them into the crystal structure factor of Equation (2.5). The atomic scattering factor of Cs and Cl is f_{Cs} and f_{Cl}, respectively. Then, the crystal structure factor is obtained as follows.

$$F(h\ k\ l) = f_{\mathrm{Cs}} + f_{\mathrm{Cl}} \cdot \exp\{\pi i(h+k+l)\} \tag{2.6}$$

In the Equation (2.6), if $h+k+l$ is an even number, $\exp\{\pi i(h+k+l)\}$ becomes 1. On the other hands, if $h+k+l$ is an odd number, $exp\{\pi i(h+k+l)\}$ becomes -1. Therefore, the following extinction rule is obtained.

$$h+k+l = \text{even number}: \quad F(h\ k\ l) = f_{\mathrm{Cs}} + f_{\mathrm{Cl}} \tag{2.7a}$$

$$h+k+l = \text{odd number}: \quad F(h\ k\ l) = f_{\mathrm{Cs}} - f_{\mathrm{Cl}} \tag{2.7b}$$

That is, the phase of Cs and Cl is the same phase and the scattering amplitude strengthen with $f_{\mathrm{Cs}} + f_{\mathrm{Cl}}$ when $h+k+l$ is an even number. On the other hands, the phase of Cs and Cl is an antiphase and the scattering amplitude weaken with $f_{\mathrm{Cs}} - f_{\mathrm{Cl}}$ when $h+k+l$ is an odd number. The intensity is proportional to the square of the crystal structure factor, and the extinction rule can be understood.

Let's calculate the crystal structure factor of α-Fe for a BCC structure using the same method of CsCl. In this case, Cs and Cl in CsCl in unit cell are replaced by Fe atom. The crystal structure factor is given as:

$$h+k+l = \text{even number}: \quad F(h\ k\ l) = f_{\mathrm{Fe}} + f_{\mathrm{Fe}} = 2f_{\mathrm{Fe}} \qquad (2.8a)$$

$$h+k+l = \text{odd number}: \quad F(h\ k\ l) = f_{\mathrm{Fe}} - f_{\mathrm{Fe}} = 0 \qquad (2.8b)$$

It is understood from the Equation (2.8a) and (2.8b) the extinction rule for BCC crystal that the Bragg reflections of $h+k+l$ with an odd number do not appear.

The above results showed that the crystal structure factor can be calculated by Equation (2.5) when we have the information on atom positions in the unit cell: diffraction pattern reproduced by calculation can be compared with the observation. But the reverse process such as to determine the crystal structure from the diffraction pattern observed is quite different. The experimental result is not the crystal structure factor itself. The difficulty is in the following points: The quantities that we can obtain by X-ray diffraction experiment are the square of the crystal structure factors that are complex number. Thus, it is difficult problem to find out the phases of structure factors which are all complex number, even if we found the absolute value of some of crystal structure factors. Unless we find the phase of the structure factor on the basis of several ideas and find atomic positions in the unit cell, we cannot say that the crystal structure was clearly solved. To clarify the crystal structure of unknown material is not necessarily easy unlike the identification procedure.

2.5 Reproducibility of diffraction pattern

In the previous section, we found that the information of crystal system[6)] and lattice constant in the sample is obtained by X-ray diffraction experiment within an allowed range of diffraction angle. In the section, we discuss that the relative change of the integrated intensities among the Bragg reflections is closely related with the crystal structure on an atomic scale. Here, we will calculate the integral intensities using well known atom positions of CsCl, based on the Equation (2.4), and we will examine to see its reproducibility of the diffraction pattern observed.

The necessary variables for calculating Equation (2.4) are diffraction angle 2θ(obs), $(\sin\theta/\lambda)$, and Miller indices $h\ k\ l$ for each diffraction lines observed. We can easily obtain the numerical values of atomic scattering factors, f_{Cs} and f_{Cl} at the $(\sin\theta/\lambda)$ from the table. By using all the data obtained we can go further to calculate the crystal structure factor by using

Table 2.2 Comparison between calculation and observation for integral intensity of CsCl: extinction rule is confirmed. For comparison between J_{obs} and J_{cal}, J_{cal} is normalized in the way that J_{cal} for (1 0 0) reflection is 1000.

No.	2θ(deg)	J_{obs}	hkl	$h^2+k^2+l^2$	$\sin\theta/\lambda$	f_{Cs}	f_{Cl}	F	m	(L_P)	$J_{\mathrm{cal}} = m\|F\|^2 (L_P)$	$J_{\mathrm{obs}}/J_{\mathrm{cal}}$
1	21.534	265.6	100	1	0.121262	49.2	14.6	34.7	6	47.2	342.4	0.78
2	30.676	1000	110	2	0.171695	45.8	12.9	58.7	12	23.6	1000.0	1.00
3	37.772	145.9	111	3	0.210105	43.2	11.7	31.5	8	15.9	128.9	1.13
4	43.903	123.8	200	4	0.242648	41.2	10.8	52.0	6	12.0	197.4	0.63
5	49.398	129.4	210	5	0.271228	39.5	10.1	29.3	24	9.6	203.4	0.64
6	54.507	336.4	211	6	0.297242	38.0	9.6	47.7	24	8.1	448.3	0.75
7	63.857	82.0	220	8	0.343284	35.7	8.9	44.6	12	6.2	149.7	0.55
8	68.230	60.4	300	9	0.364052	34.8	8.6	26.2	6	5.5	116.4	0.52
			221		0.364052	34.8	8.6	26.2	24	5.5		
9	72.478	89.2	310	10	0.383719	33.9	8.4	42.3	24	5.0	219.8	0.41
10	76.618	26.8	311	11	0.402379	33.2	8.2	25.0	24	4.6	70.8	0.38
11	80.685	30.7	222	12	0.420199	32.4	8.0	40.4	8	4.3	57.2	0.54
12	84.718	20.7	320	13	0.437347	31.7	7.8	23.9	24	4.0	56.4	0.37
13	88.724	112.1	321	14	0.453844	31.1	7.7	38.8	48	3.8	279.5	0.40
14	96.747	10.4	400	16	0.485197	29.9	7.4	37.4	6	3.5	29.5	0.35
15	100.787	27.2	410	17	0.500094	29.4	7.3	22.1	24	3.3	79.5	0.34
			322		0.500094	29.4	7.3	22.1	24	3.3		
16	104.889	46.9	411	18	0.514583	28.9	7.2	36.1	24	3.2	154.5	0.30
			330		0.514583	28.9	7.2	36.1	12	3.2		
17	109.061	13.6	331	19	0.528644	28.4	7.1	21.3	24	3.2	35.0	0.39
18	113.364	31.4	420	20	0.542411	27.9	7.0	34.9	24	3.1	92.9	0.34
19	117.802	20.6	421	21	0.555809	27.5	6.9	20.6	48	3.1	64.0	0.32

Equation (2.5). Since the crystal structure has been given by the Equation (2.7a) and (2.7b), submitting numerical values to the Equation (2.7a) and (2.7b), we can calculate F_{hkl} of each reflection peaks. In Table 2.2, the multiplicity m_{hkl} and Lorenz-polarization Lp factor are given. The integrated intensities J_{hkl}(calc) were calculated by multiplying the correction factors to the values obtained after taking square of $|F_{hkl}|$. All the calculated values of J_{calc} listed in the Table 2.2 are normalized for $J_{\text{calc}}(110)$ to be 1000 as seen. The ratio $J_{hkl}(\text{obs})/J_{hkl}(\text{calc})$ are also listed in the last line of the Table 2.2, which should be constant, being independent of the reflections.

Figure 2.5 (a) shows the F_{hkl}(calc) as a function of 2θ. The figure shows a change of $f_{\text{Cs}} + f_{\text{Cl}}$ and $f_{\text{Cs}} - f_{\text{Cl}}$ with diffraction angle, namely it shows the extinction rule. In Figure 2.5 (b), the $J_{hkl}(\text{obs})/J_{hkl}(\text{calc})$ are plotted against each diffraction angle of 2θ. If the calculations are in agreement with observations, all the values should be 1, because both the calculation and observation are normalized at the 1 1 0 Bragg reflection as seen in the Table 2.2. This fact shows that the calculation is getting to becoming systematically large with the increase in the diffraction angle, as shown by guideline. It suggests that there need another correction factor something like a damping factor to reduce the calculations. If you measure the sample again at an elevated temperature, something like a 100 °C. or more higher temperature, you will have a result showing that the damping become more remarkable. Those facts suggest that the effect of thermal vibration on the crystal structure factor needs to be considered. About an effect of thermal vibration, see the next section.

2.6 Effects of thermal vibration

In the Equation (2.5) in calculating the crystal structure factor, we expressed the position of the j-th atom in a unit cell by $\boldsymbol{r}_j(x_j, y_j, z_j)$. The crystal which is made of atoms in a material at room temperature, however, are in a state of thermal vibration. The crystal lattice is fluctuating with time so that every atoms and molecules are considered to be vibrating around their equilibrium positions with time. Let such thermal vibration of atoms and molecules, however, be independent of the location of the unit cell in a crystal. This is called one body approximation or Einstein approximation in the treatment of thermal vibration of material. Within a harmonic approximation of the thermal vibration of an atom the average of the square of the **atomic displacement** $\langle \boldsymbol{u}_j{}^2 \rangle$ is proved to be proportional to the absolute temperature T in a form of $k_{\mathrm{B}}T$, where k_{B} is the Boltzmann's constant. Thus, the square of the atomic displacement $\langle \boldsymbol{u}_j{}^2 \rangle$ should be understood to be different depending on the measurement temperature. In the calculation of the structure factor, we did not completly take into account this fact.

Although all the atoms in a crystal are always displaced with time and also according to their location in the crystal lattice, with lattice vibration, it is possible to define the equilibrium position of the j-atom in the unit cell. Let it be $\langle \boldsymbol{r}_j \rangle$ as the result of taking statistical average. In taking the statistical average, there are two ways to take it. One is to take the average of the displacement within a certain limited space and the other is to take its average in a certain time. These physical phenomena are often distinguished by saying the time average and the space average. However, we should say that it is difficult to distinguish the two cases by the diffraction experiment. The displacement of the atom due to thermal vibration is time dependent so that let it be $\boldsymbol{u}_j(t)$. Then, the atomic position is written by $\boldsymbol{r}_j(t) = \langle \boldsymbol{r}_j \rangle + \boldsymbol{u}_j(t)$.

The variation of time in an atom is the order of pico seconds, if consider the lattice vibration. This value is negligibly small compared with the time that need in the diffraction experiment. Also it is impossible to distinguish an atom in a specific unit cell in crystal by X-ray diffraction experiment,

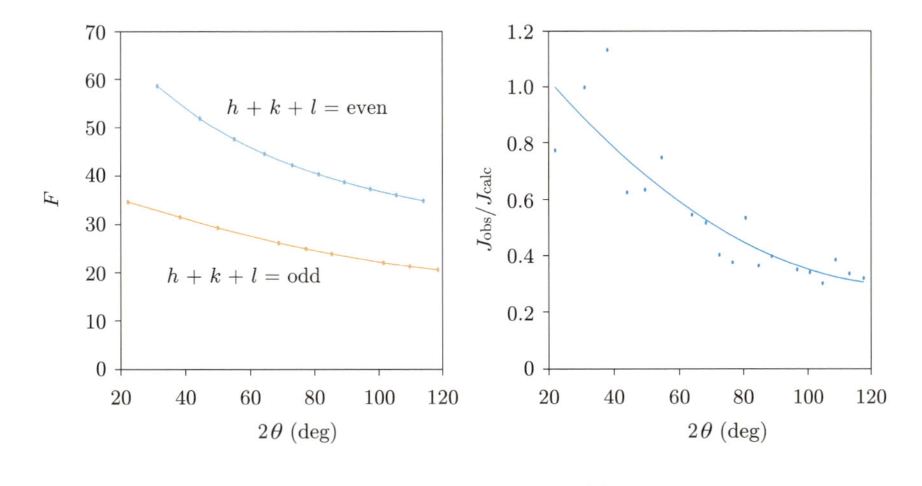

(a) Change of crystal structure factor F_{hkl} with the diffraction angle 2θ

(b) Change of $J_{hkl}(\mathrm{obs})/J_{hkl}(\mathrm{calc})$ with the diffraction angle 2θ

Figure 2.5 (a) Crystal structure factor F_{hkl} as a function of diffraction angle 2θ, (b) $J_{hkl}(\mathrm{obs})/J_{hkl}(\mathrm{calc})$ as a function of diffraction angle 2θ.
The ratio $J(\mathrm{obs})/J(\mathrm{calc})$ decreases systematically from 1 with the increase of 2θ, due to the fact that thermal vibration of atom is ignored, as explained in text.

since the coherent region of one photon can be considered to be more than the size of a crystallite. This means that the value obtained by our diffraction experiment is not only the statistical time average but also the space average. Now, the average crystal structure factor is written in the following formula in which ⟨- - -⟩ means to take a statistical average.

$$\begin{aligned}&\langle F_{hkl}(\sin\theta/\lambda)\rangle\\&\quad=\sum_j f_j\left(\sin\theta/\lambda\right)T_j\left(\sin\theta/\lambda\right)\exp\{2\pi i(hx_j+y_j+lz_j)\}\end{aligned}\tag{2.9}$$

where, $T_j(\sin\theta/\lambda)$ is the temperature factor and is often called **Debye-Waller factor**. It is allowed to regard a correction factor to atomic scattering factor $f_j(\sin\theta/\lambda)$ due to thermal vibration. If assume that the amplitude of atomic thermal vibration is the same for all the directions, which is nothing but the isotropic approximation for thermal vibration, $T_j(\sin\theta/\lambda)$ is proved to be converted into the following simple formula.

$$T_j(\sin\theta/\lambda) = \exp\left\{-8\pi^2\langle u_j{}^2\rangle(\sin\theta/\lambda)^2\right\} \tag{2.10}$$

$$8\pi^2\langle u_j{}^2\rangle = B_j \tag{2.11}$$

where, $\langle \boldsymbol{u}_j^2\rangle$ is the thermal average of the square of the displacement of the j-th atom. Equation (2.10) can be easily interpreted by using a simple consideration based on the thermal motion of a particle of mass m which is placed in a harmonic potential given by quadratic curve. Its physical meaning is shown by Figure 2.6 (a). As the result, the thermal average of $\langle \boldsymbol{u}_j^2\rangle$ is expressed to be proportional to the temperature (T: absolute temperature) as $k_{\mathrm{B}}T$. The average value multiplied by $8\pi^2$ is expressed as B_j which is called **temperature parameter** or **thermal parameter**. The temperature parameter is one of the parameters which should be determined (in conjunction with atom position) from the crystal structure analysis.

Equation (2.10) is the Gauss function with respect to $\sin\theta/\lambda$. This is a simple decrement function telling that its value is 1 at $\sin\theta/\lambda = 0$ and decreases with increasing $\sin\theta/\lambda$. We see that the experimental result, shown in Figure 2.5 (b), is qualitatively well in agreement with the Equation (2.10). The illustration of this phenomenon in real space is Figure 2.6 (a) and (b). Figure (a) shows that an atom with charge distribution $\rho(r)$ is located in a harmonic potential given by the quadratic curve, and Figure (b) describes a moment of the atom in thermal vibration in the potential. You will see that the charge density $\rho(r)$ is spread out something like $\rho'(r)$.

In general, B_j takes a large value to the crystal of a low melting point but a small value to the crystal of a high melting point. If B_j is analyzed as an abnormally large or small value, you should think that there are any other reasons. In the most cases, its origin is in a fact that the diffraction pattern itself obtained by the measurement is not correct. Thus, it should

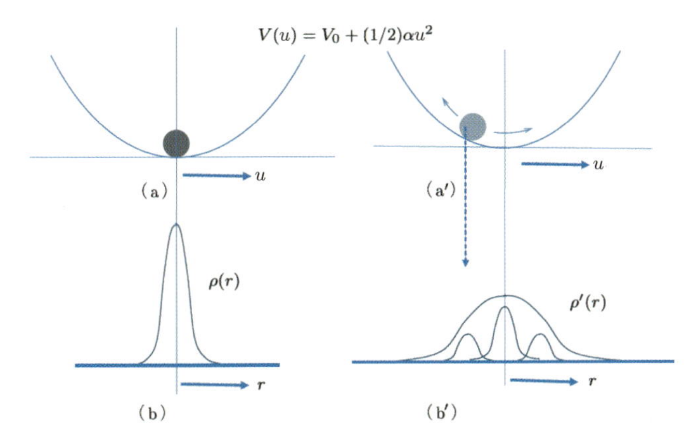

ρ(r): election density distribution of an atom

ρ'(r): effective electron density distribution of an atom spread by thermal vibration

u : displacement of an atom

r : coordination of the electron density

Figure 2.6 Illustration of change of electron density distribution between stationary atom and thermally vibrated one.

(a) An atom being at the center of potential represented by a quadratic curve

(a′) An atom being one moment on a shoulder of the potential curve

(b) Electron density distribution of an atom

(b′) Expanded electron density distribution by thermal vibration

be careful in obtaining the B_j-value by the structure analysis.

The Gauss function of Equation (2.10) is of an isotropic function, because the equation is derived with an assumption of an isotropic thermal vibration of atom. However, atoms in a crystal are not necessarily located in a spherically symmetric potential. There are many cases we have to consider atoms with anisotropic thermal parameter, depending on the crystal symmetry. This problem is not included in this book. There is a commentary[28] about the treatment of anharmonic potential.

Annotations

1) The powder X-ray diffraction pattern/line seems to be named as the Debye-Scherrer diffraction pattern/line in a European textbook. These words are used with the Japanese textbooks. In an American textbook, however, such an expression seems to have not been used even in the report of ICDD. Instead the phrase of powder diffraction pattern/line is seen. Because it seemed oddly, the author opened "Dictionary of physics" published (1984) by Baifukan in Japan and checked this. It is listed as follows: Ring-formed X-ray diffraction pattern is observed from a powder sample which is called as Debye-Scherrer rings in Europe, since Debye and Scherrer found it in 1916. However, such a diffracton pattern had been found already by Shoji Nishikawa and Suminosuke Ono in 1913 in Japan. Similar circumstances are thought to have been also in the United States. The author learned this word from a well known book "Crystalline State -1 General Survey-" written by W. H. Bragg and W. L. Bragg. In addition, some of the senior scientists had used this Debye-Scherrer pattern/rings, then the author has just followed them in using it without any doubt until now. For such a reason, the author is paying particular attention to use this word in the present textbook. It seems to be confusion of the times when the communication, transmission of the information, was not as good as today.

2) In order to simplify a calculation, let both the wavelength of X-rays and the lattice spacing d be 1 Å, where 1 Å $= 0.1$ nm. The Equation (2.1) becomes $\theta = \sin^{-1}\{1/2\}$. The Bragg angle θ becomes 30 degrees, and the reflection angle 2θ is 60 degrees. Therefore, it is recommended to select the wavelength of X-rays close to the lattice spacing of your sample crystal. This is for making easy to observing the diffraction.

3) In the case of a rough crystallite of whose surface is not smooth, as shown in Figure 1.7 (c), the number of atoms M which form the atomic plane must be different in every lattice plane so that it is not a constant number. When we consider the diffraction of crystallite with such rough surfaces by introducing the concept of statistics, you should be understand in such a way that the mean value of M contributes to the scattering.

4) The integrated intensity formula shown for the diffraction line is based on the **kinematical diffraction theory**. Kinematic diffraction theory is on an assumption that the X-rays which are diffracted once by a crystal in a sample get out from the sample without diffraction again in the sample. In general, this assumption is appropriate when the probability that diffraction happens is very low. You would get some idea about it by picking up a simplified example. Let assume that a diffraction phenomenon is caused with 10 % of probability, as for the probability that diffraction gets up successively twice, it becomes 1 %, as the probability is calculated as $(1/10) \times (1/10)$. The size of crystallites of powder samples is considered to be less than several μm. This is the size that is unnecessary for us to take account of such a multiple diffraction phenomenon, or a multiple scattering, except a few diffraction lines of which intensity is strong. If the size of the crystallite is big with high quality, however, most of incident X-rays cause a Bragg

reflection in the crystal and the X-rays reflected become the incident X-rays which cause Bragg reflection again. Between the incident X-rays and the reflected X-rays interference phenomenon happens, which cannot be interpreted by the kinematic diffraction theory. The dynamical diffraction theory is the theory dealt with this phenomenon exactly as possible. This phenomenon is not included in this book.

5) $\Delta(x)$ is called delta-function that is defined as that it takes 1 only when the variable x is equal to 0, otherwise 0 for any value of x.

$$\begin{cases} \Delta(x) = 1 & \text{at} \quad x = 0 \\ \Delta(x) = 0 & \text{at} \quad x \neq 0 \end{cases}$$

If this function is used to $x = \sin\theta/\lambda - 1/2d_{hkl}$, it is meaningful only when Bragg condition $x = 0$ is satisfied. the function is equivalent to say that the diffraction angle 2θ is satisfied with $\sin\theta/\lambda = 1/2d_{hkl}$ on the lattice plane h, k, l. Therefore, $\Delta(\sin\theta/\lambda - 1/2d_{hkl})$ is the function representing Bragg condition.

6) In the previous chapter, the samples of the cubic crystal system were selected because it is easier to discuss, and our discussion has been forwarded. How to determinate crystal system from the diffraction pattern is another important subject, although have not taken.

Chapter 3

Identification of crystalline material

The crystal structures of many materials have already been elucidated by many researchers, and the obtained data have been registered to several organizations and those each data was compiled as databases in the organizations. Therefore, we are now in a situation that we can analyze any materials easily by comparing the diffraction data obtained from the sample with the database. From the process, we understand the information that what kind of crystal structure forms the material and what kind of impurity is included in. This operation is called the identification by X-ray diffraction. In this chapter, its operation process will be explained by showing the usefulness of the present technique in the evaluation of the quality of given materials.

3.1 Identification

Qualitative analysis of a material by X-ray diffraction method is the technique to select the diffraction pattern that is corresponding to the pattern obtained from the sample among the existing diffraction patterns. Qualitative analysis in powder X-ray diffraction is, therefore, called as **identification**. This is a fairly unique phrase in the field of analytical science. The reliability of the result is dependent on the quality of the database and also on the technique to obtain high quality X-ray diffraction patterns from the sample.

As for the database, **ICDD** file (card)[2], [12] is widely accepted. ICDD is the International Centre for Diffraction Data. This organization was established in 1941. It started from the database of the manual search that is called the Hanawalt method at the beginning. According to the sales catalog (2013 – 2014), it says possible to provides the program of PDF-2 (Powder Diffraction File-2), PDF-4+, WebPDF- 4+, PDF- 4/Minerals, and PDF- 4/Organics. However, not only ICDD database but also the databases provided by other organizations are available. As an example, we can cite the utilization of **COD** (Crystallography Open Database). Besides, if own database has been ready by calculation in advance when the structures of the materials which are frequently quoted are well known, it is possible to identify the result obtained by experiment in comparison with the calculations. This is nothing but the establishment of own database.

According to a recent ICDD catalog, PDF-2 Release 2013 is mainly the database of inorganic materials, and this database was compiled with the cooperation of **FIZ** (Fachinformationszentrum Karlsruhe) and **NIST** (National Institute of Standards and Technology). It is described that the number of data sets is 265,127 including the well-known organic crystal. The database contains the information on the spacing (d-value), relative intensity, Miller indices, chemical formula, chemical name, mineral name, structural formula, space group, melting point, and density and so on. The number of data sets of PDF-4+2013, which contains programs for the qualitative analysis as well as the identification, is 340, 653, including WebPDF-4+ which is able to search with high speed using internet. The number of data sets of PDF-4/Minerals and that of PDF-4/Organics 2014 is 40,424 and 479,278, respectively. PDF-4/Organics includes the program of the identification to explore an unknown data, including the atom coordinate in unit cells, including a program of the Rietveld analysis[1)] so as to be able to perform qualitative analysis.

In order to identify the data that obtained by using MiniFlex 300/600, the program PDXL developed by RIGAKU for powder data analysis is useful. Here, the method of identification will be explained based on the ICDD database, and then a characteristic of PDXL is supplementary shown.

3.2 Contents of ICDD cards

For instance, the ICDD card of Al_2O_3 is shown in Figure 3.1. The number 00-042-1468 at the top of the card indicates the card number of Al_2O_3. Next to the number, quality of the data is given. S, I, C, O, and B are the marks indicating the quality of the data. The quality of S means the highest reliability data. I means that, although the intensity of diffraction has been reexamined, the quality is lower than S. C is the data obtained by calculation. O is the data of low reliability, and B is the data which has not been evaluated. The chemical formula and the name are written as Al_2O_3 and Alumina. The next column shows the crystal system, space group, lattice parameter, lattice volume, references, and the diffraction pattern. In the bottom, it shows intensity data, 2θ, d-value, and relative intensity I/I_{100}, where the reflection of which the intensity is represented by (1 0 0) in the table is denoted by I_{100} and the intensities of other reflection are given by the relative intensity I/I_{100}. Finally, the Miller indices h k l are represented.

3.3 Search based on ICDD cards

The first step in identification is to find the d-value of which reflection has the highest intensity. The d-values are used to classify materials into 45-groups, and are arranged in the order of the group with the largest d-value to the group with the smallest d-value group. Next, within the selected group, we compare the d-values and their intensity ratios for three reflections. After selecting a number of candidates, we compare the d-value and intensity ratio of the observed reflection in detail to those of the candidates to identify the sample material. As the d-value has 3 – 4 effective digits, the precision of $0.05° – 0.10°$ is sufficient in reading of the Bragg angle. The intensity at the peak position is enough for the comparison of intensities. Because the **integrated intensity** is almost proportional to the intensity, if FWHM (Full Width at Half Maximum) of the peaks are the same. As an example of a searched result in an ICDD card, the detailed procedure of a search based on the Figure 3.1 is explained as follows.

(1) Pre-processing (peak search)

For qualitative analysis, before performing a peak search, we have to eliminate the background and smooth the data. This is done by entering processing conditions to the computer. In doing this, we must set an accurate position for the background and avoid excessive smoothing. To locate and identify trace components, we must also avoid dropping small peaks and exaggerating other peaks.

(2) Primary search

We perform a primary search to list candidate compounds from the ICDD file on the hard disk. Since this search examines several tens of thousands of ICDD cards, specifying the appropriate search conditions is critical. This means specifying the ICDD card file (including sub-files), contained elements, and so forth. The search will return compounds meeting the search conditions entered here. The **search file** and **error windows** are explained in detail below.

① Search file : Select the inorganic materials file if the sample is an inorganic compound and the organic materials file if the sample is an organic compound. In addition, there are subfiles listed in Table 3.1.

② Error window : The interplannar spacing shown may not perfectly match the standard data, if the sample stage is set at a non centrosymmetric position rather than at the standard position or if a sample is utilized which contains some element as an impurity and is in a state of solid solution. Thus, the observed peak position should be given a margin to allow judgment of a match with standard data. Refer Chapter 7 in the X-ray optics. The extent of this margin is called the error window.

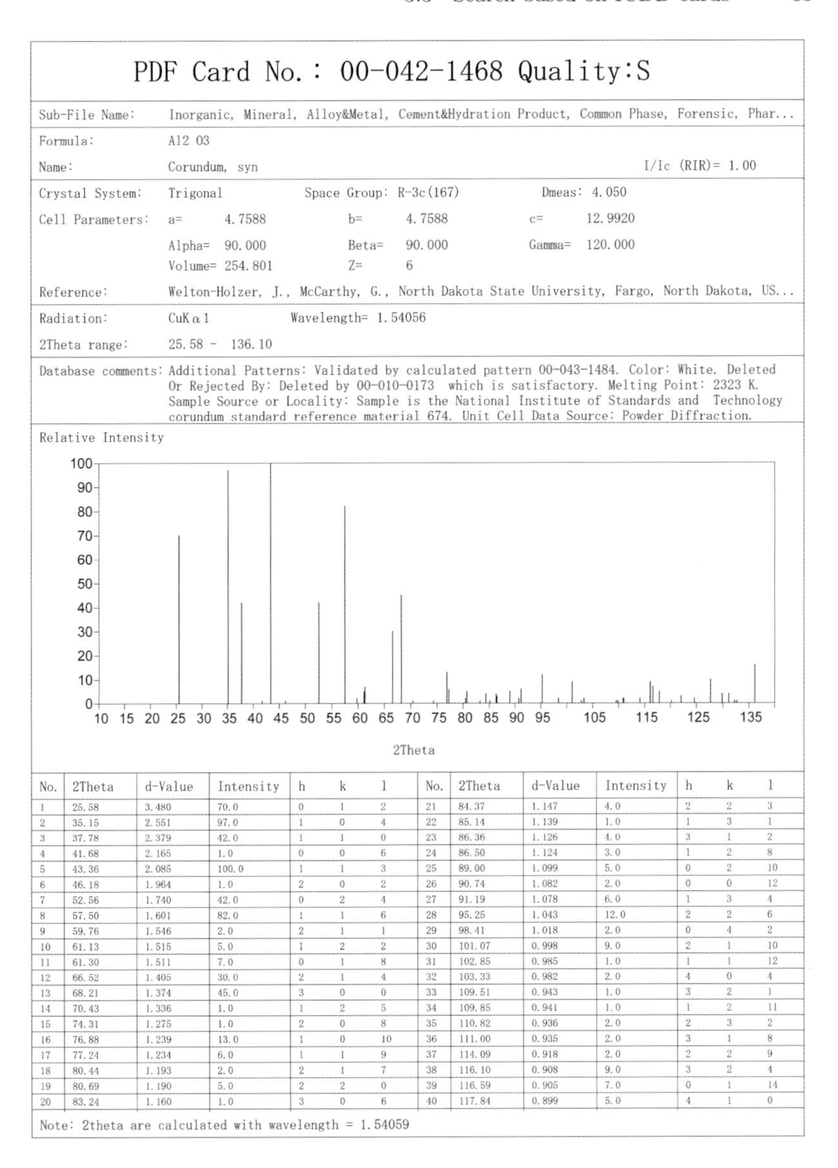

PDF Card No.: 00-042-1468 Quality:S

Sub-File Name: Inorganic, Mineral, Alloy&Metal, Cement&Hydration Product, Common Phase, Forensic, Phar...

Formula: Al2 O3

Name: Corundum, syn I/Ic (RIR)= 1.00

Crystal System: Trigonal Space Group: R-3c(167) Dmeas: 4.050

Cell Parameters: a= 4.7588 b= 4.7588 c= 12.9920

Alpha= 90.000 Beta= 90.000 Gamma= 120.000

Volume= 254.801 Z= 6

Reference: Welton-Holzer, J., McCarthy, G., North Dakota State University, Fargo, North Dakota, US...

Radiation: CuKα1 Wavelength= 1.54056

2Theta range: 25.58 - 136.10

Database comments: Additional Patterns: Validated by calculated pattern 00-043-1484. Color: White. Deleted Or Rejected By: Deleted by 00-010-0173 which is satisfactory. Melting Point: 2323 K. Sample Source or Locality: Sample is the National Institute of Standards and Technology corundum standard reference material 674. Unit Cell Data Source: Powder Diffraction.

No.	2Theta	d-Value	Intensity	h	k	l	No.	2Theta	d-Value	Intensity	h	k	l
1	25.58	3.480	70.0	0	1	2	21	84.37	1.147	4.0	2	2	3
2	35.15	2.551	97.0	1	0	4	22	85.14	1.139	1.0	1	3	1
3	37.78	2.379	42.0	1	1	0	23	86.36	1.126	4.0	3	1	2
4	41.68	2.165	1.0	0	0	6	24	86.50	1.124	3.0	1	2	8
5	43.36	2.085	100.0	1	1	3	25	89.00	1.099	5.0	0	2	10
6	46.18	1.964	1.0	2	0	2	26	90.74	1.082	2.0	0	0	12
7	52.56	1.740	42.0	0	2	4	27	91.19	1.078	6.0	1	3	4
8	57.50	1.601	82.0	1	1	6	28	95.25	1.043	12.0	2	2	6
9	59.76	1.546	2.0	2	1	1	29	98.41	1.018	2.0	0	4	2
10	61.13	1.515	5.0	1	2	2	30	101.07	0.998	9.0	2	1	10
11	61.30	1.511	7.0	0	1	8	31	102.85	0.985	1.0	1	1	12
12	66.52	1.405	30.0	2	1	4	32	103.33	0.982	2.0	4	0	4
13	68.21	1.374	45.0	3	0	0	33	109.51	0.943	1.0	3	2	1
14	70.43	1.336	1.0	1	2	5	34	109.85	0.941	1.0	1	2	11
15	74.31	1.275	1.0	2	0	8	35	110.82	0.936	2.0	2	3	2
16	76.88	1.239	13.0	1	0	10	36	111.00	0.935	2.0	3	1	8
17	77.24	1.234	6.0	1	1	9	37	114.09	0.918	2.0	2	2	9
18	80.44	1.193	2.0	2	1	7	38	116.10	0.908	9.0	3	2	4
19	80.69	1.190	5.0	2	2	0	39	116.59	0.905	7.0	0	1	14
20	83.24	1.160	1.0	3	0	6	40	117.84	0.899	5.0	4	1	0

Note: 2theta are calculated with wavelength = 1.54059

Figure 3.1 ICDD card of Al_2O_3.

Table 3.1 Subfile name for ICDD check.

Inorganic	Common phase	Ionic conductors	ICSD pattern
Organic	Corrosion	NBS	Cambridge pattern
Mineral related	Educational pattern	Pharmaceutical	NIST pattern
Alloys Battery	Gunpowder	Pigments	Linus Pauling pattern
Cement	Forensic	Polymer	history
Ceramic	Giant magnetoresistance	Super conducting material	
	Intercalate	Zeolite	

(3) Second search

The standard data for the compounds in the primary search results is superimposed on patterns obtained from the sample on the PC screen. (Compare the *d*-value and relative intensity in the entire diffraction profile obtained against the standard data to check for correspondence.) If any of the standard data overlaps the measured profile almost exactly, the sample can be said to be identified. If you come across unidentified diffraction lines, identify them in the same way. Repeat this identification process until no diffraction lines remain to be examined.

As an example, Figure 3.2 shows the data of a sample, a mixture of anatase and rutile, obtained with the MiniFlex 300/600. The observed diffraction pattern shows black line. The results of anatase and rutile searched in ICDD card are drawn as the red line and blue lines, respectively. The data for these materials explains all observed reflections. The FOM (Figure of Merit) written in the search result is the reliability to the observed data. It is dependent on the degree of agreement of *d*-values obtained, relative intensity, and the number of peaks and so on with database. The larger the coefficient, the more reliable the analysis. Keep in mind that this value is a benchmark. The final judgment must be made by humans. Let's see how closely the angle and intensity in the obtained data match

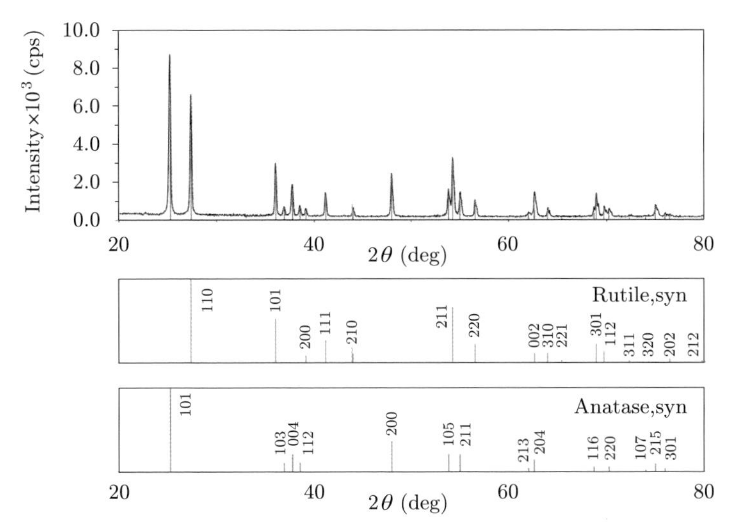

Figure 3.2 X-ray diffraction pattern of TiO_2 mixture of anatase and rutile and the identification result by ICDD search. Black line is the profile obtained. Red and blue line are the ICDD result of the anatase and the rutile, respectively.

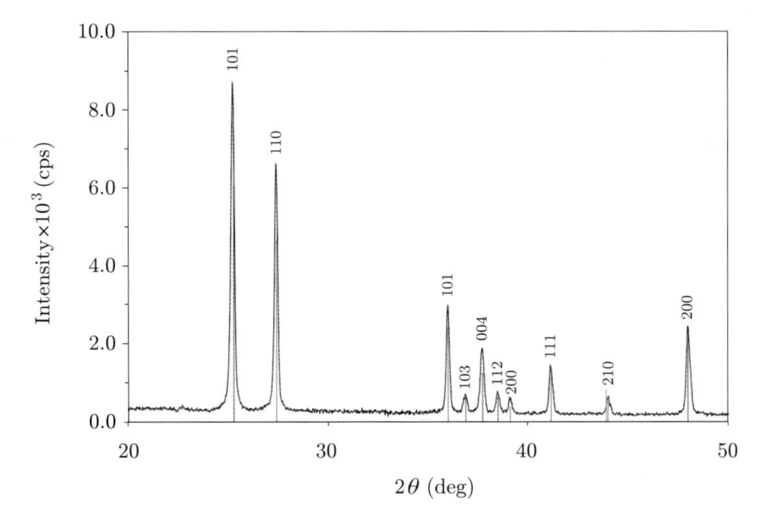

Figure 3.3 Comparison between observation and ICDD result. Red and blue lines are of anatase and rutile obtained from ICDD result.

the data registered in the ICDD file. Figure 3.3 shows the results of a comparison to anatase. The degree of correspondence here is acceptable.

3.4 Comments on identification process

During the identification process, the *d*-value and I/I_{100} of the diffraction lines from the sample may differ slightly from the ICDD data. The factors leading to this discrepancy fall into three categories: (1) causes originating from an improper adjustment of equipment and sample; (2) causes stemming from characteristics specific to the sample; and (3) causes attributable to some doubtful reliability of the ICDD data.

(1) Device adjustment and sample preparation

① Needless to say, the equipment, MiniFlex 300/600, must be serviced and maintained. If the goniometer's zero point deviates by an angle $\Delta 2\theta$, for example, all diffraction lines will shift by $\Delta 2\theta$. Such errors are due to mechanical misalignments.

② In the results of measurements based on the B-B method, the peak profile becomes asymmetric. With $2\theta = 90^\circ$ at the center, reflections appearing on the low-angle side have tail on the low-angle side, while reflections appearing on the high-angle side have tail on the high-angle side. If this umbrella effect is questionable, reduce the divergence angle of the incident optical system in the vertical direction to decrease the effect.

③ In the case of the measurement using Kβ filter, diffraction pattern with the Kβ line remains with the high intensity peaks. This problem can be avoided by placing an analyzing crystal (monochromator) in front of the detector.

④ Most diffractometers, including the MiniFlex 300/600, do not have sufficient resolution to distinguish between the Kα_1 and Kα_2 contained in CuKα X-rays. If the average size of crystallites in the sample is relatively large and the quality of crystallites is high, diffraction lines of Kα_1 and Kα_2 may separate on the high-angle side and appear as independent diffraction lines. Again, unexpected must be noted.

⑤ If the X-ray tube used is old and the surface of the target is contaminated, characteristic X-rays attributable to impurities in the contamination may mix. If the filament material, tungsten (W), adheres to the target, WLα (= 1.476 Å) rays will appear. As the detected wavelength will be between CuKα rays and CuKβ rays, you must be vigilant.

⑥ Note that the observed data might also include diffraction lines by the sample holder and equipment cover.

(2) Characteristics specific to the sample

In the X-ray diffraction by powder sample, the diffraction intensity ratio I/I_{100} happens to change depending on the sample preparation or characteristics specific to the sample. It will be bringing to abnormalities in the profile of the diffracted X-rays, thereby impeding proper identification of the sample. However, an investigation of these causes involves the analysis of the state of the sample. Actual cases are listed below as reference examples.

① In the powder sample with large particle size (approx. 10 μm or larger), the cone-shaped diffraction from powder sample (see Figure 2.1) shows irregularities in its intensity distribution. Recommend to make it sure by seeing Figure 8.3. In certain cases, relative intensity I/I_{100} can vary by several tens of percent. We can minimize this effect by grinding the sample so that the grain diameter is less than approximately 10 μm (until particles cannot be felt by a fingertip) or by rotating the sample around the axis perpendicular to the sample surface. In MiniFlex 300/600, a rotary sample holder has been prepared.

② If the grain diameter is 0.1 μm or smaller, on the other hand, the diffraction lines become wider and tend to overlap on the high-angle side, making identification difficult.

③ In a powder sample with needle-shaped crystallites or flat-plate-shaped crystallites, certain diffraction lines (for example, diffraction line from the (0 0 1) plane in the case of crystallites that are elongated along the c-axis) will exhibit sharp edges along the axis of width, but diffraction line in the perpendicular direction will expand in width.

④ The crystal grains in the sample may be oriented preferably. Relative intensity can vary significantly, depending on the degree of preferred orientation. In extreme cases, we observe reflections of only specific indices. This can occur in crystal samples with a layered structure, such as clay minerals and graphite powder, and in foils and fibrous materials. This can also be a problem specific to a certain sample.

⑤ In certain cases, a material of the same structure with a slightly different lattice constant may be mixed in. This happens when the impurities remain in a sample so that the sample is in a state of **solid solution**. For example, the lattice constant of $BaTiO_3$ is $a_0 = 3.97$ Å. However, a solid solution of $SrTiO_3$ with $a_0 = 3.91$ Å will result in locally irregular composition, causing the diffraction lines to separate. In certain cases, the shape of the unit cell may be slightly distorted, degrading symmetry. This can be seen in materials with a perovskite structure (such as $BaTiO_3$) and intermetallic compounds. In these cases, the profile of the diffraction line becomes noticeably asymmetrical. For samples with **imperfect solid solutions** of CaO in MnO, the profiles exhibit tails on the high-angle side. These problems can also be specific to a certain sample.

⑥ In samples (e.g., graphite, talc, $Mg_3Si_4O_{10}(OH)_2$ and so on) with **stacking faults**, the profile of the diffracted X-rays from an irregular surface will exhibit a tail on the high-angle side.

Above six items are the examples that the peculiar properties of the sample disturb the identification of the sample. To locate the cause is equivalent to evaluate the polycrystalline state of the sample (**crystalline**

texture), this is important as another challenge.

(3) Reliability of ICDD data

① Much of the data obtained before 1950 are based on measurements obtained by the photographic method, which is sometimes associated with the results of unreliable analyse. Although old data have been replaced with new data as new data becomes available, the ICDD data might still contain older, unreliable data.

② If you measure an unknown material with a wavelength different from the data indicated on the ICDD card, relative intensity may differ from ICDD data. It is recommended to use the appropriate X-rays.

3.5 PDXL software (RIGAKU)

In identifying an unknown X-ray diffraction data, a search method has been described by using a search program based on a database, both of which are provided by ICDD, in the previous section. This search method has been explained as if there is no other method. Because this is the method which has been the most commonly used for a long time. During these ten several years the studies in analyzing and evaluating the crystal structure as well as the studies on the analysis of the polycrystalline state of material have been rapidly progressed on the basis of the X-ray diffraction data obtained. The programs of analysis became surprisingly enriched through those studies. Since the free database prepared by Japanese Society of Crystallography is available to use now in Japan, it may be one idea to use such a program together in your identification of an unknown X-ray diffraction data.

So far, the study of the structure analysis has been understood as the determination of the atomic positions and its thermal parameters in the unit cell within the error. The material science was able to discuss on an atom scale on the basis of the results obtained by X-ray structural study. Since the Maximum Entropy Method (**MEM**) [2] was introduced, the electronic distribution such as bonding electron to link two atoms in unit cells became visualized three-dimensionally. Thereafter, it has developed to be able to discuss material science not only on an atomic scale but also on a scale of electron distribution. In the case of neutron diffraction, the information obtained by MEM is not charge density distribution but the distribution of atomic nucleus so that the anisotropic potential on which atom is located can be discussed.

On the other hand, the profile of the Bragg reflection to be provided by a powder X-ray diffraction contains a lot of information, but it did not attract attention too much. The reason is as following. The profile which gives the resolution of equipment was found to be reproducible using a function called Pearson VII. However, the relations between this function and the parameters which describe X-rays optics of the instrument have not fully been understood. The problem solved very recently.

If the parameters reproducing X-ray optics of device are known, it turns out that the proper profile of the device which we call the resolution function can be reproduced. It has become possible to analyze a characteristic profile due to the sample in detail. Specifically, the size distribution of crystallites in the sample, the average shape, and distortion of crystallites are analyzable. In addition, such a physically meaningful quantity is obtained to describe the state of materials as orientation of the crystallites in sample. The analytical program that can support the request of researchers is enriched. The search program can adopt novel analytical technique developed in these studies. Since the MEM method and profile analysis were introduced as described above, it is understood that a progress has been made in the powder X-ray diffraction method which have never been before.

The company RIGAKU provides a program for the data analysis called PDXL and is planning the improvement in this program. This software contains a program for an above-mentioned crystal structure analysis and also a program aiming at the analysis of crystalline texture in materials based on the profile analysis. Hope to use it. Six options are prepared for qualitative analysis, quantitative analysis, applied analysis, Rietveld refinement, structure analysis, and cluster analysis. The program has the following three advantages.

① Narrowing a search range by selecting basically eight diffraction lines with relatively high intensities and also setting a condition, it is possible to search the result with high reliability (with high figure of merit).

② The profile which represents the resolution of the device became possible to be reproduced by a computer. Thus, the peculiar problems with respect to the device may be solved. Therefore, the inherent property of samples, which has been mentioned in Chapter 3.4 (2), can be analyzed. For instance, although the identification of the sample of which crystallites are oriented is one of the difficult analyses, it became possible to identify with fairly high reliability. It is said to be possible to analysis even in the case that the average lattice parameter shows characteristic anisotropy which is due to the anisotropic distortion of crystallites. In powder X-ray diffraction method, it is an innovative program that was to be able to take advantage of the state analysis of the material.

③ In the structure analysis of an unknown crystal, there are famous indexing programs that have been paid for or can be copied legally by anyone: DICVOL, ITO, and N-TREOR[23]. In addition, Rietveld Refinement[24] program which reproduce the whole powder diffraction pattern by a least squares fitting procedure may be used to determine the lattice parameters and the structure of unknown sample. Furthermore, if Maximum Entropy method (MEM)[3] is utilized, the electron density distribution in the unit cell can be reproduced. Therefore, it is possible to challenge the structure analysis of a new material by using a powder diffraction pattern.

3.6 Analysis based on characteristic diffraction lines

When performing a qualitative analysis of powdered material that generate characteristic diffraction pattern, such as clay minerals[2], [8] and rock minerals[9], we need to focus on peculiar diffraction pattern to determine the materials included. Shown below is an example.

When X-rays of the CuKα wavelength are used to measure a sample of α-quartz (SiO_2) powder, we observe strong diffractions at diffraction angles (2θ) of 20.7° and 26.5°, and also a quintuplet pattern around the angle of 68°. Figure 3.4 shows the profile of the quintuplet. This diffraction pattern is unique to α-quartz and is not found with other materials. If we see this diffraction pattern from our sample, we can presume that the sample contains α-quartz. In the case of α-quartz, however, the intensity of the quintuplet is weak, thus, we need to perform a careful inspection.

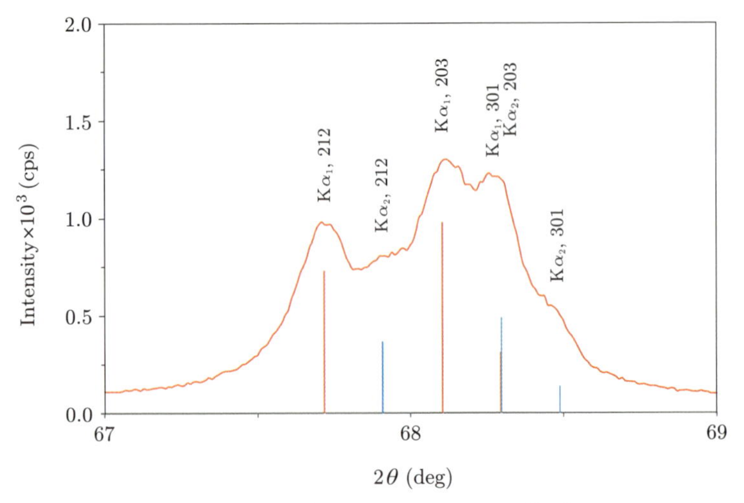

Figure 3.4 Profile of quintuplet (SiO_2).

3.7 Identification of organic compounds

Despite the abundance of organic compounds, there are fewer registered ICDD cards for organic compounds than for inorganic compounds. This means the systematic identification technique used for inorganic compounds such as clay minerals, rock minerals, and metals is less suitable for the identification of organic compounds. Since organic compounds can be characterized by unique molecular shapes, their diffraction patterns are often complex. Nonetheless, if the diffracted X-rays of homologous compounds and allied compounds of the organic compound to be identified are available, we can perform qualitative analysis (identification) in the same way as for inorganic compounds.

Annotations

1) Rietveld refinement method was devised by Hugo Rietveld who was a researcher in the Netherlands, in 1969. The method was introduced for the analysis of the neutron scattering data of powder samples. The analytical method used at that time to solve the crystal structure was a least-squares method, which refines the parameters to describe the atomic positions in a unit cell, on only the basis of the integrated intensities of observed diffraction lines. Paying attention to the fact that the profile of the diffraction line obtained by neutron diffraction can be well reproduced by relatively simple Gaussian function, Rietveld proposed a least squares fitting procedure to reproduce all the diffraction pattern by adjusting lattice parameters as well as structural parameters of sample. Because diffraction angle 2θ of the diffraction line has to be taken account in reproducing whole diffraction pattern, this method has an advantage that not only the atom position but also the lattice constant can be estimated by the least square refinement method. Hence, this method was also called as the total fitting method. It was inherited to Alan Hewatt (later Grenoble Institute) who stayed in Harwell Research Institute (Atomic Energy Research Establishment, AERE). With participation of David Cox in the Brookhaven National Laboratory, a program seemed to be improved afterwards. This technique developed in the field of neutron diffraction has been popularized immediately. It would be happening to think also by anybody. Its application to X-rays diffraction was immediately tried by many researchers, although it did not go well. Because the profile of the diffraction line obtained by X-ray diffraction was not reproduced by simple function. After ten several years later, Dr. Toraya, who became the director of X-ray Research Institute of RIGAKU by living his professor post of Nagoya Institute

of Technology, found an appropriate form of function to give a key to the solution. The improvement of the program has been examined by many scientists and continues up to the present day. Regardless of X-rays or neutron, the program commonly used in Japan would be RIETAN, which was developed by Dr. Izumi of National Institute for Materials Science[24]. However, this program is not popular in any countries other than Japan.

2) MEM is an abbreviation for the Maximum Entropy Method. We have mentioned that the crystal structure factor is the quantity obtained by Fourier transform of electron density distribution in a unit cell and it is generally a complex number. The intensity of a diffraction line is proportional to the square of the crystal structure factor. Thus, if taken root square of the intensity it is proportional quantity to the absolute value of the structure factor. But, its phase is not obtained though regrettable. Strictly speaking, it should say that the structure is not decided uniquely. As described in the text, therefore, in structure analysis, a least square refinement method is utilized so as to refine structure parameters such as atomic coordinates and thermal parameters until the diffraction intensities calculated on the basis of those parameters are in good agreement with those observed. In mathematics, there is a problem which is called "the inverse problem" by finite number of experimental data. Let a function be B(K) which is obtained by the integral transformation of a function A(X). The inverse problem is to seek the original function A(X) from B(K), which is, however, not given all of them. Determination of crystal structure from the diffraction data is nothing but this inverse problem. One of the way to solve this problem is to use Maximum Entropy Method. First of all entropy has to be defined which is a quantity corresponding to the entropy that is known in physics. Then, above our subject can be reduced to a problem in order to reproduce the electron density in a unit cell under a certain restraint referring to the data obtained from diffraction experiment by maximizing the entropy defined above. As the result it became possible to reproduce the electron density distribution that can be said not to be more ambiguous. At present, is evaluated as the most effective method, it is used.

In Japan, Prof. Y. Hosoya of the Institute for Solid State Physics, Tokyo University reported his study about usefulness of MEM for the phase determination of the crystal structure factor in Annual Meeting of Physical Society of Japan at the first half of the 1970s. Although the present author was not aware the progress of his study since then, after ten several years, Dr. Steve Wilkins of the CSIRO of Australian stayed in the author's laboratory and brought his idea of the MEM. A fairly long discussion had been made about what sort of informations we could get from MEM. It turned out that there were two possibilities in applying MEM. One was the phase determination of structure factor and the other was the reproducibility of electron density distribution in unit cell. At that time, Dr. Makoto Sakata, who was one of staffs, went to Steve's laboratory of

Australia and stayed there for three months. After coming back to Japan, Makoto got a prospect that could reproduce electron density distribution by the collaboration with one of the postgraduates.

Chapter 4

Texture of crystallites in material

The X-ray diffraction pattern obtained from a given sample is characterized by a series of Bragg reflections. When those reflection angles and the integrated intensities are analyzed, the average crystal structure of crystallites is determined on an atomic scale. The crystallites are the components constituting the sample material. This chapter discusses that the state of crystallites in a material can also be determined by analyzing the profile of the diffraction line.

4.1 Crystalline state and amorphous state

Depending on temperature and pressure, the substance under investigation will be in one of the three states: vapor, liquid or solid. In physics, the word "phase" is often used instead of "state", such as liquid phase or solid phase. Besides, there are two solid states: amorphous state and crystalline state. The structural differences between these states are shown in Figure 4.1, which are given by W.H. Zachariasen[4] in 1932. Both materials have the same chemical symbol, SiO_2. Many texts have reproduced these diagrams, showing the structural difference between the two solid states. Quartz is in a crystalline state; the Si-atoms (small violet circles) are covalently bonded via oxygen atoms (grey circles) in a periodic arrangement. However, silica glass lacks periodic regularity in their arrangement. The amorphous state was initially thought to be unique to glass, but has since been identified as well in pure silicon and in certain alloys. Methods for manufacturing these substances in the amorphous state now exist because of various applications that require its superior uniformity of this state. The preceding addressed the existence of two morphologies of solids.

Figure 4.2 shows the X-ray diffraction patterns of quartz and silica glass obtained by MiniFlex 300/600. The differences between two states are obvious. The sharp diffraction peaks of the quartz are easily enable the indexing. The identification of silica glass by diffraction experiment is also easy by confirming a broad peak seen near the scattering angle of $2\theta \fallingdotseq 28°$.

Let's observe the diffraction pattern of the liquid phase in a low melting temperature material. (In MiniFlex 300/600, high- and low-temperature sample stage that can be easily installed is available. However, for the present purpose it is not effective, because the sample stage will be inclined during the measurement by the $\theta - 2\theta$ scanning.) The diffraction pattern of liquid state is almost the same as that from the amorphous state. From this result, we infer that samples in the liquid and amorphous states have similar microstructure. Thus, two states cannot be distinguished by X-ray diffraction. The difference between the two states is fluidity. In the liquid

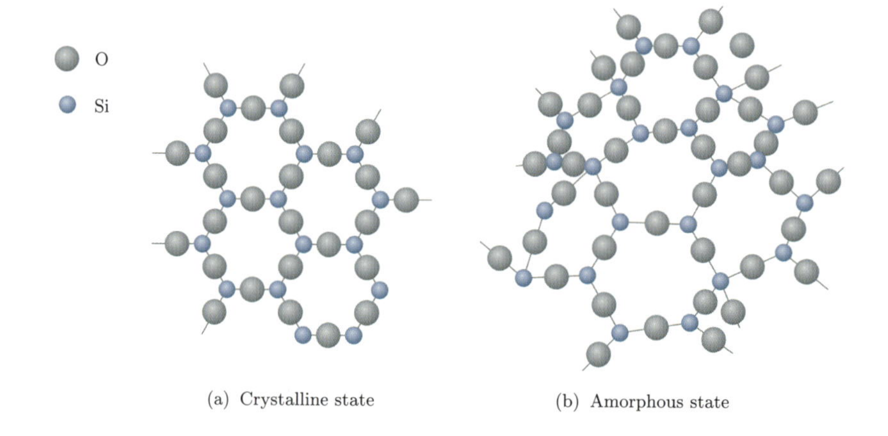

(a) Crystalline state (b) Amorphous state

Figure 4.1 Two solid states of SiO_2.

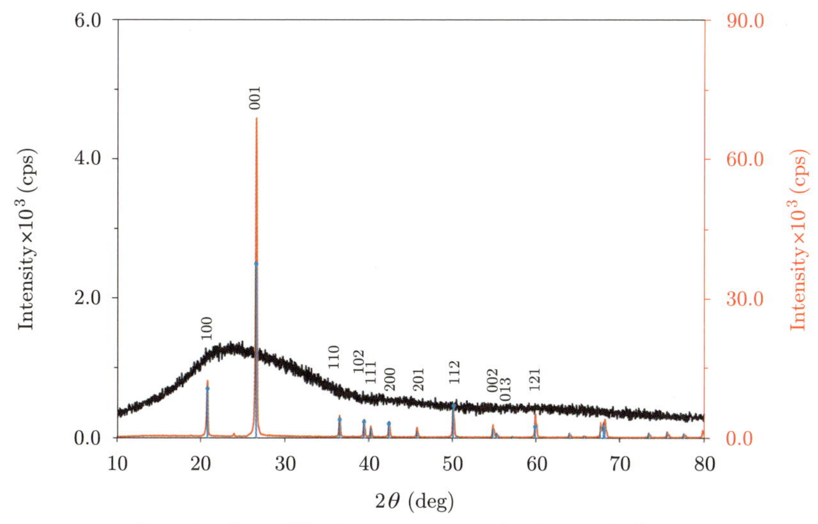

Figure 4.2 X-ray diffraction patterns of powdered quartz (red) and SiO2 glass (black). The blue lines are peak positions reported by ICDD.

state, atoms and molecules are in a state of flux, whereas in the amorphous state, their fluidity is extremely low. However, even with such extremely low fluidity, plastic deformation will occur due to the fluidity of atoms and molecules in this configuration, if we apply counteracting forces to silica glass for an extended period. The key aspect here is that the crystalline and amorphous states occur at ordinary temperatures and pressure, although their microstructures are quite different at first glance. In the crystalline state, atoms and molecules are arranged periodically. In the amorphous state, they are arranged randomly, keeping interatomic distances to the closest atoms. According to solid-state physics, two states have almost the same free energy.

Here, the SiO_2 glass is used to illustrate the amorphous phase. The structure of the polymer resin, which is referred to as "plastic", contains a locally crystallized region. Moreover, its overall structure is also amorphous.

4.2 Crystalline state

When crystalline state is classified according to X-rays diffraction method, six states are sorted as shown in Figure 4.3. First of all, it is divided into two groups; the **single crystalline state** and the poly-crystalline state. The materials discussed in this booklet are in a polycrystalline state. On the other hand, the single crystalline state is that you can imagine a single crystallite in a polycrystalline material grown up to big size. It is the same state as that of the natural jewels, quartz, emerald, and diamond. Single crystal state can be further divided into two states according to the X-ray diffraction method: that is the perfect single crystals and the mosaic single crystals. Although the main topic here is how to characterize materials in the polycrystalline state, we here deseribed briefly the evaluation method of single crystal state shown in Figure 4.3.

The quality of the single crystal is also evaluated through observing the width of a Bragg reflection by using the X-ray diffraction. If explain it a little in detail, it is necessary to measure the width of the Bragg reflection in a high precision of 1/400°. This observation method is called **rocking curve** measurement. If the sample has minimal lattice defects and the crystals feature an ideal periodic structure, the **full width at half maximum (FWHM)** will be an angle of less than a dozen arcseconds. As for the measurement of the Bragg reflection with a high degree of accuracy, you need to prepare a special single-crystal diffraction apparatus[6]. In the Figure 4.4 (a), the principle of the measurement is shown. Using a monochromator consisting of one-crystal and 2-crystals (channel-cut), X-rays emanating from the X-ray source should be monochromated so that Kα_1 and Kα_2 are well separated, and incident beam is well parallel. We put a single crystal sample on the sample stage of the diffractometer, of which surface is parallel to the lattice plane from which Bragg reflection we intend to measure. The measurement of Bragg reflection is usually made

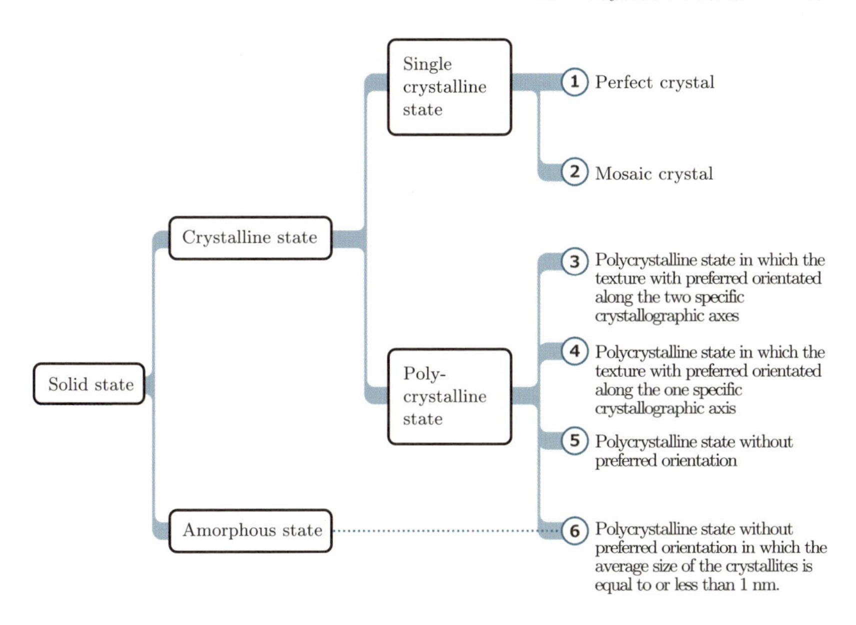

Figure 4.3 States of solid state in material.

symmetrical condition so that the incident X-ray beam is set to be θ angle with the surface, and the counter arm is set to be 2θ angle with the incident beam direction. After this setting, the both the incident beam and the counter arm are fixed so as not to move. During the measurement, we do not put slit in front of the counter, saying such a condition "wide open". The measurment is performed by changing θ from low angle side to the high angle side of the Bragg angle. In this way, a sample stage only is rotated by a very small step width which is less than $1/200°$, during the measurement. It looks like that only the crystal is rocking during the measurement, after fixing all other parts of the diffractometer. Thus, this measurement is called the rocking curve measurement. In the diffractometer for the measurement of single crystal sample, sample holder is designed to be able to rotated by θ, independently with 2θ rotation of counter arm.

Thus, in order to avoid confusion 2θ-rotation and θ-rotation, the axis of this sample rotation is formally called ω-axis and the scanning of sample is called ω-scanning.

The examples obtained by such an instrument are shown in Figure 4.4 (b) for the three samples: Si and LiF single crystals, and Pyrolytic Graphite. In the measurements, we utilized an instrument "SmartLab" by installing the incident optics shown in Figure 4.4 (a). This diffractometor was developed by RIGAKU. as a fully automatically controlled intelligent X-ray diffractometer and its sample stage can be kept to be always horizontal. The width of the rocking curve of Si single crystal is observed within $8.8''$ which is less than ten minutes. In general, the width depends on the quality of crystal. There is a theoretical formula[5] representing the width of locking curve based on the dynamical diffraction theory, which is applicable if the crystal is ideally perfect. If the observed width is very close to the calculation, we may say that the crystal under the measurement is almost the perfect crystal, which corresponds to "perfect crystal ①" in the Figure 4.3. In the Figure 4.4 (b), the perfect crystal is only Si crystal. The crystals known as ideal single crystal at present are Si, Ge, quartz, and ice. In recent, there is SiC, which is a compound semiconductor and is attracting attention in industry, because of its high melting point. The variety of its growth method for high quality single crystal is examined. Generally, if there exist lattice defects in a crystal, the degree of perfectness decreases with the increasing defects in the crystal. Accordingly, the width of rocking curve becomes broad. The width of the rocking curve in the most crystals is within 0.1 – 0.3 degree, being far from its perfect state. The single crystals showing such width are called the **mosaic crystal** in the Figure 4.3. In the field of crystal growth, the quality of the crystal which has just grown is now evaluated by comparing the width of the rocking curve with its calculation.

On the other hand, we say that the sample showing the powder diffrac-

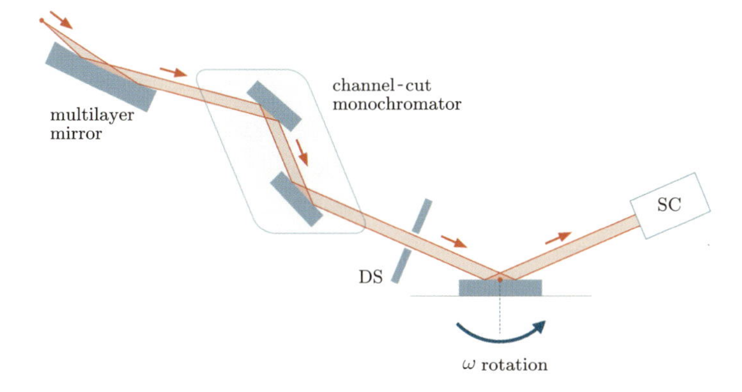

(a) X-rays optics for rocking curve measurement of single crystal

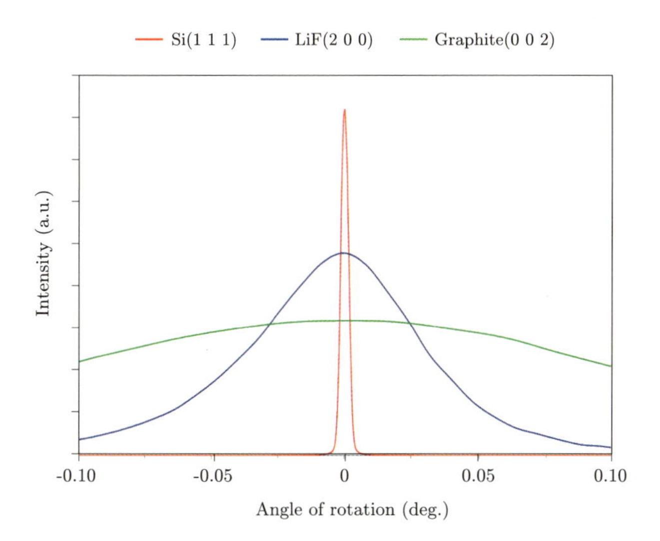

(b) Rocking curve of Si, LiF, and PG,
FWHM of Si, LiF, and PG is 8.8″, 3.0′, and 0.53°, respectively.

Figure 4.4 X-rays optics for rocking curve measurement of single crystal and the obtained profiles.

tion pattern is in the polycrystalline state. Metals and ceramics, which are able to obtain anywhere, are corresponding to the materials of the polycrystalline state. When we look its surface through an optical microscope after polishing well the surface of a metal chip such as iron and aluminum, and removing the portion processed deformed by etchant, we can observe an aggregate of particles with different shapes and sizes[1)] which are less than several μm. With further observation of the particles with an electron microscope such as SEM and TEM, there is often that the particles are one of the crystal. But, you can also see that the particles are consisted of several small crystals, which are called the **crystallites**. Suppose that the diffraction pattern of such crystallites is observed via a two-dimensional detector, such as a photographic film. If the diffraction rings obtained are so uniform, we say that the polycrystalline state of sample is in a state of ⑤ of Figure 4.3 (See Figure 2.2 and Figure 4.7 (a)).

On the contrary, there is a case in which the diffraction rings contain several spots or intensified regions. This fact is indicating that a lot of crystallites that forms the sample are arranged toward a particular direction. This is the **texture with preferred orientation**. Among them, polycrystalline state in which the texture with preferred orientation can be classified by two states. One is that crystallites are oriented along the two specific crystallographic axes (corresponding to ③ of the Figure 4.3). The other is the polycrystalline state in which crystallites are oriented along the one specific crystallographic axis (corresponding to ④ of the Figure 4.3). This is a classification based on X-ray diffraction pattern.

There are more or less differences in the degree of orientation in crystallites of real sample material, although the most of samples can be regarded to be in the state of orientation. The ideal polycrystalline state without preferred orientation can be said to be an extremely rare occasion (corresponding to ⑤ of the Figure 4.3). As mentioned before, the powder X-ray diffractometer can only measure a part of the diffraction rings. Accordingly, in the measurement of the sample that consisted of the texture with preferred orientation, the proper intensity change among diffraction lines is not reflected definitely. It, therefore, disrupts the identification of the sample. In order to overcome the problem, several methods to improve are recommended in the manual of X-ray diffractometer. Powder sample with uniform particle size should be prepared or sample stage should be rotated around an axis perpendicular to the sample surface.

Finally, the case is of a polycrystalline state which corresponds to ⑥ of the Figure 4.3. Please try to imagine a state in which the average size of the crystallites is equal to or less than 10 nanometers (corresponding to the ⑤ of the Figure 4.3). When the size of the crystallite is decreases, the ratio of the number of the atoms on the surface to the total number of atoms is significantly increased. For instance, if the size of crystallite becomes 1 nm, its ratio will be 0.6. It means that the most of constituent atoms belong to a part of the surface. Thus, the lattice planes are difficult to be confirmed by diffraction experiment. The width of powder diffraction ring becomes broader and broader, and finally will be overlapped with neighbor diffraction rings. The pattern might be resembling the diffraction pattern of amorphous; the pattern is close to the ⑥ of the Figure 4.3.

Once again, consider the state of the ④ based on the state of the ⑤ in the Figure 4.3. The polycrystalline state of the ⑤ is that the crystallites are distributed around its average size. In addition, the crystallites are distributed randomly: it is the state without preferred orientation. As for the polycrystalline sample with preferred orientation, the crystallites are arranged along a specific crystallographic axis, irrespective of its size distribution. Such a polycrystalline sample is said to be in the state of the **textured orientation** or the **preferred orientation** along one of the crystallographic axis. In other axes which are perpendicular to the oriented axis, the crystallites have to be in the state without preferred orientation. This state corresponds to the ④ of the Figure 4.3, and the

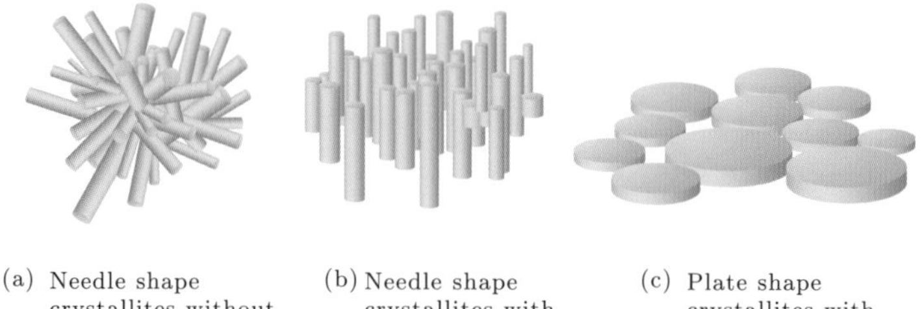

(a) Needle shape crystallites without preferred orientation (b) Needle shape crystallites with preferred orientation (c) Plate shape crystallites with preferred orientation

Figure 4.5 Three examples of poly-crystalline states.

drawings (b) and (c) in the Figure 4.5 represent those states. These states exist in fiber texture, such as the cold drawn metal wire and the metal materials rolled under a certain temperature. As will be shown in the subsequent paragraphs, the crystallites in those cases are of a plate like or a needle like shapes. Because it shows the diffraction pattern is significantly anisotropic. It is a very good example for X-ray diffraction method to be useful, especially in the microscopic (the size in the area of μm) structural evaluation of the material.

In the poly-crystalline sample, there is a case that the crystallites are arranged along two crystallographic axes. It is the state of the ③ of the Figure 4.3. When we look at salt, every salt particles are observed has a dice shape. We say that the **crystal habit** is seen. There is a case that every particles has a dice shape and is oriented towards two crystalline axes, although the crystal-habit is not a major cause of two orientation axis. It is called the state with bi-axial orientation. In this case, it is very common that the sizes of crystallites are distributed around an average size. It can be however considered as one of a state in a crystal with many lattice defects, so that the orientation of crystal lattice is locally different and changes here and there. We can say this is the mosaic state of a single crystal and corresponds to the ② of the Figure 4.3. By using a high-resolution X-ray diffractometer for single crystal, as seen in the Figure 4.4, it is possible to evaluate it and also to distinguish those states.

If look at materials which are in the solid state from a microscopic point of view, you will see that it is classified into many crystalline states. Nowaday, various type of X-ray diffractometers have been already developed exclusively for the evaluation of perfect single crystals and also in that for any type of mosaic crystals. If you managed the powder X-ray diffractometer, you may surely obtain various information by the evaluation of crystalline texture. A little more specific stories will be given in the following.

4.3 Influence of crystalline state on diffraction line

Various information about crystallites in sample are included in the diffraction lines. Here, we discuss the relation between the profile of the diffraction line and the crystallites. In general, observed powder X-ray diffraction pattern consists of several Bragg reflections which are given by scattering or diffraction from various crystallites with a variety of size. For convenience, we here consider the profile of the Bragg reflection from one crystallite, which involves a detailed review of Section 2.2. Although it is what you have already understood, the relations of d_{hkl}, $2\theta_{hkl}$, the peak intensity, and the width of the Bragg reflection are as follows. ① If the spacing d_{hkl} is the same under the condition of the same wavelength, the diffraction angle $2\theta_{hkl}$ which causes the Bragg reflection is the same, regardless of the size of crystallite. ② if the crystallite has the N number of sheets in the lattice plane, the peak intensity of the Bragg reflection is proportional to the square of the N, N^2. In addition, ③ the width B_{hkl} of the Bragg reflection is inversely proportional to the number of lattice plane$N$: it is directly proportional to $1/N$. The intensity distribution of such a diffraction line shows in the Figure 2.4. Mathematically, it is a function of

$\sin^2 Nx/\sin^2 x$. This function is called the diffraction function or also called the **Laue function**. Here, $x = 2\pi d_{hkl}\sin\theta/\lambda$ is the scattering angle. As shown in Figure 2.4, the value of x is an integer at an angle where Bragg reflection appears. This function shows a peak when x takes the integer. Since it was not so easy to evaluate the FWHM, B_{hkl}, of Laue function, P. Scherrer[7] derived the approximation to replace the Laue function by a Gaussian: $\sin^2 Nx/\sin^2 x \rightarrow N^2 exp\{-(Nx)^2/\pi\}$. In this approximation we have:

$$B_{hkl} - b = 0.94\lambda/L_{hkl}\cos\theta \qquad (4.1)$$

In this equation, b is often called the natural line width, which expresses the instrumental resolution. Often b is neglected because it is very small compared with B_{hkl}. If this is not the case, b should be subtracted from the observed width B_{hkl}. As seen in Equation (4.1), the observed width is inversely proportional to the size of the crystallite, $L_{hkl} = Nd_{hkl}$. The unit of the width is in radians and the crystallite size is in Å. Although this equation is derived for the width of the scattered or diffracted X-rays from a single crystallite, Scherrer was applied it to analyze a broad diffraction pattern from a gold colloid. The colloid was assumed to be composed of small cubic crystallites. The colloid size obtained from the equation was reported to be 19 – 86 Å.

Thus, although Equation (4.1) is the fairly rigorous equation to determine the size of uniformly distributed crystallites, its applicability to a general powder X-ray diffraction ring is debatable. However, the **Scherrer equation**[7] is a convenient expression that gives physical quantity to represent the size of crystallites.

4.4 Average size of crystallites

Sample measured by MiniFlex 300/600 is always in powder form. The intensity distribution of the diffraction line observed (line profile) is considered to be the overlapping of the diffraction lines from plenty of crystallites in the powder sample. In view of the intensity distribution, thus, the application of Equation (4.1) and its interpretation need to reconsider.

When the X-rays are incident on the sample, there are also a plenty of crystallites with the same orientation which satisfying the Bragg condition in the region exposed by X-rays. All the X-rays scattered from different crystallites are incoherent to one another, because of their irregular arrangement. That is to say, the profile is observed as the result of the overlapping of several Laue function due to crystallites with different sizes. Let the Laue function of the crystallites consisted of N lattice plane be $L_N(x)$. Furthermore let the proportion of crystallites with N lattice plane is P_N, which is normaized by $\Sigma_N P_N = 1$. By using P_N, the profile of the observed diffraction line becomes $I(x) = \Sigma_N P_N L_N(x)$. By the same way of the Scherrer equation, Laue function $L_N(x)$ is replaced with the Gaussian function $G_N(x)$. Thus, the observed profile is given by $I(x) = \Sigma_N P_N G_N(x)$.

In any way, the crystalline size L_{hkl} can be found by the Scherrer equation if one can estimate the line width B_{hkl} from the profile $I(x)$ of the observed powder diffraction line. This corresponds to approximation to replace the profile $I(x)$ by one Gaussian function. Strictly speaking, therefore, the obtained L_{hkl} is not guaranteed as the average size of the crystallites. Because no theoretical and experimental studies related to the above argument are available in the literature, a characterization study[25] of Pd nanoparticles prepared by using a special technique called VEROS is cited here. Figure 4.6 is the results shown by the study. The three techniques were used in the estimation of the average size of Pd-nanoparticles. The first technique is the direct observation of the distribution of the particle size by electron microscope, shown by histogram. The average size is given as $L_{EM} = 22 \pm 3$ Å. The size distribution of crystallites can also be estimated by Fourier analysis of the profile of a Bragg reflection, although

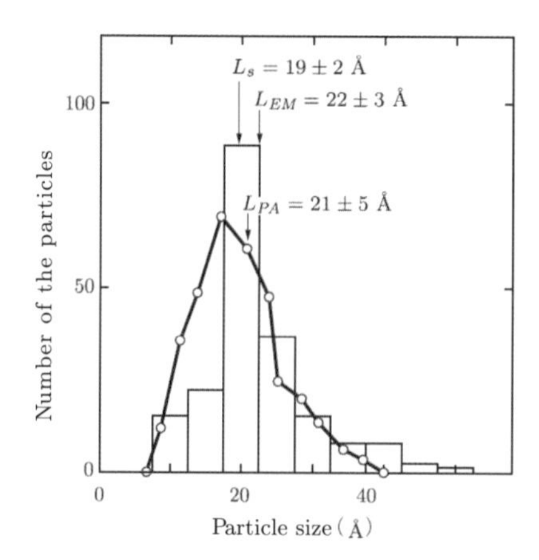

Figure 4.6 Particle size distribution of Pd ultrafine particles.

its technique is not given in this textbook. The result obtained by such an analysis using the (2 2 0) Bragg profile is shown by line graph and $L_{PA} = 21 \pm 5$ Å is estimated. The average size obtained on the basis of the Scherrer's equation by using the (2 2 0) Bragg reflection is $L_S = 19 \pm 2$ Å. All these three results are in good agreement within the error estimated.

In addition, according to the author's experiment of the sample in which crystallites are of slightly bigger average size, there was not remarkable difference between the average size obtained by the Scherrer equation and the average value of the distribution function obtained by Fourier analysis. Thus, it may be said that the Scherrer equation gives an index of the average size of crystallites in a sample.

Let us investigate the range of crystalline size. Let's input 1.54 Å of CuKα into λ and 45 degree to θ in the Equation (4.1). The diffraction angle 2θ is 90 degree. The width of the diffraction line, $B_{hkl}(2\theta)$, is assumed to be 0.5 degree (= 0.00875 rad). Then, we have the average size of crystallites to be 230 Å. Conversely, the line width was obtained to be 0.1 degree, when the average size of crystallites was 1100 Å under the same diffraction condition.

It means that the resolution of the diffractometer should be under 1/5; the step width becomes about 2/100 degree. This is an appropriate step for the MiniFlex 300/600 based on the B-B optics.

If the average size of crystallites increases beyond 1 μm = 10,000 Å, the number of crystallite which contribute to the diffraction decreases in the region exposed by X-ray beam. Then, it becomes sharp, but not uniform. These diffraction lines which are called "spotty diffraction lines". MiniFlex 300/600 is the X-ray diffraction instrument which is designed to give uniform diffraction profiles from the whole powder sample as much as possible. However, data analysis should be done with care, because the non-uniform diffraction pattern may appear, as shown in Figure 4.7 (a) – (c). The X-ray beam was incident on a part of the flat-shape powder sample and observation of diffraction pattern was paformed by a cylinder type X-ray camera. The samples were quartz powder with different crystallite sizes. You will see the change of diffraction patterns from uniform diffraction pattern (a) to a spotty one (b) and (c), as the average crystallite sizes increases.

4.5 Defects in crystallite

So far, we have regarded all the crystallites in our sample to be of good quality, having lattice planes without defects. The real crystal is, however, not perfect and contain defects of various types, which is called the **lattice defect**. For instance, an atom is missing at the lattice point: it is called the vacancy. In the case, surrounding lattices are shrinked toward the vacancy. On the contrary, an atom gets into the place that is not a lattice point: it is called the interstitial atom. The lattice interval becomes partially longer or shorter than the average value. These effects appear statistically. The existence of the defects in crystallites affects on the width of the powder diffraction line, as seen by modifying the equation of the Bragg reflection.

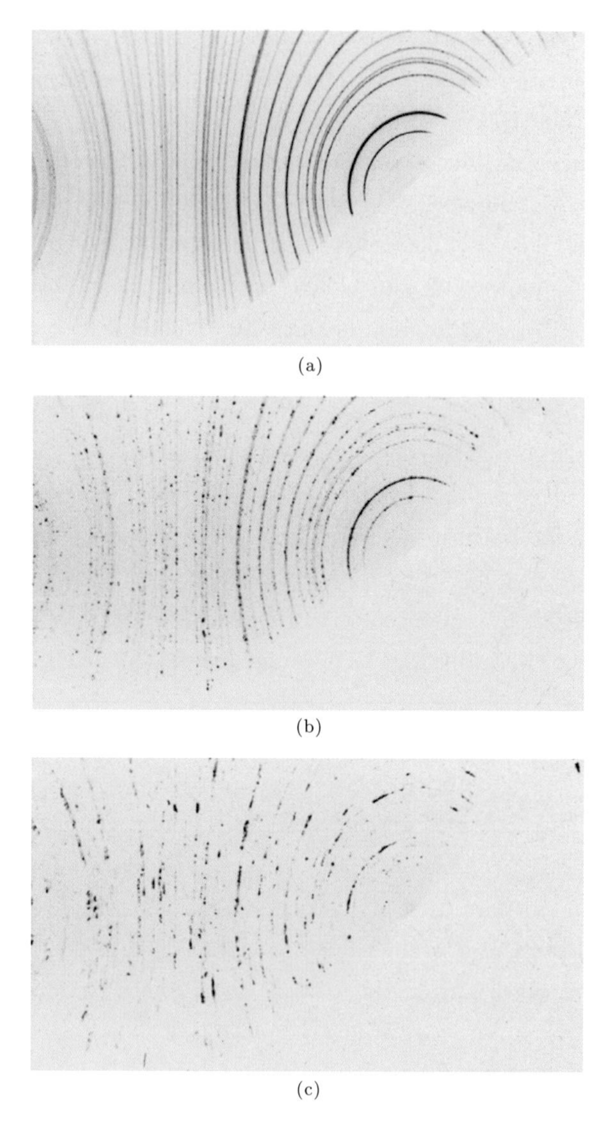

Figure 4.7 X-ray diffraction patterns of powdered quartz. The sizes of crystallites are (a) < (b) < (c). The right side lower part is not observed in the diffraction pattern due to the interruption of sample holder.

In the Bragg equation, the spacing d is distorted by Δd. By total differential of the Bragg equation, the diffraction angle 2θ is shifted by $-2\Delta\theta = 2\Delta d/d \tan\theta$. The diffraction angle will increase or decreas from the average value, depending on the expansion and the contraction of the lattice parameter. When the average distortion is $\langle \Delta d/d \rangle = \eta$, the width B'_{hkl} of the diffraction line is

$$B'_{hkl} = 2\eta \tan\theta \tag{4.2}$$

The equation indicates that the width B'_{hkl} of diffraction line broadens with increasing diffraction angle 2θ. This equation is called the Hall equation[8], since it was derived by Hall. If we consider this broadening of the width with that due to crystalline size, the observed width B^{obs}_{hkl} is given by the sum of the crystallite size B_{hkl} and the internal distortion B'_{hkl}.

$$B^{\text{obs}}_{hkl}\left(\frac{\cos\theta}{\lambda}\right) = 2\eta\left(\frac{\sin\theta}{\lambda}\right) + \frac{0.94}{L} \tag{4.3}$$

If we plot a quantity $B^{\text{obs}}_{hkl}(\cos\theta/\lambda)$ against $(\sin\theta/\lambda)$ (diffraction angle), it is expressed by a straight line. An avereage distortion η can be given from the slope of the straight line and the average size of the crystallites L(Å) is obtained from the value of B^{obs}_{hkl} at $(\sin\theta/\lambda) = 0$. Although this equation is convenient to find the average crystallite size and also the lattice distortion, the data may not be well handled in this equation. Thus, you should consider it when you analyze the data.

4.6 Strain of crystal lattice

In general, any material shrinks under compressive stress. Conversely, it stretches by applying a tensile stress. In a region of Hooke's law, however, it will return to the original, if those stresses are removed. If we extend this phenomenon to crystallites which are constituents of material on an atomic scale, it corresponds to the shrink and stretch of the crystal lattice.

It means that we may confirm it by X-ray diffraction method. If specifically added, it may be possible to estimate the amount of shrinkage or stretching of the crystallites which are oriented to a specific direction inside of the material by measuring the shifts of X-ray Bragg angles. We can examine "presence or absence of strain" by whether the systematic shifts of Bragg angles exist or not. If the elastic constants is known for the material you are dealing with, you may find what sort of stress exists in a given material as well as the stress distribution.

This method obtaining strain of crystal lattice is one of the longstanding applications of the X-rays diffraction. The basic idea behind is simple and may be understood based on the fundamental formula given below. If you know the lattice constant of the distortion-free material, you can easily calculate Bragg angle of the reference. Next, the Bragg angle of a sample with a strain is determined by directly measuring the sample. Since the difference $\varDelta\theta$ of its value from the reference is obtained, the strain can be given by the following equation.

$$\frac{\varDelta d_{hkl}}{d_{hkl}} = -\cot\theta\varDelta\theta \tag{4.4}$$

Here $\varDelta d_{hkl}/d_{hkl}$ on the left-hand side is a relative shift of the lattice spacing; it indicates the distortion ε. The equation shows that the distortion ε can be calculated directly from the amount of shift of the Bragg angle. The coefficient $\cot\theta$ in front of $\varDelta\theta$ denotes that a relative displacement $\varDelta d_{hkl}/d_{hkl}$ is significantly increased with the increase of Bragg angle θ, telling us that relative shift is more accurately determined if the Bragg reflection of a higher angle is utilized.

4.7 Preferred orientation of crystallites

Here, we describe in more detail the preferred orientations of crystallites. As a key point in identifying a given material to be of the preferred orientation, as we have already seen in the previous section; a material consisted of crystallites of a needle shape lead to the following interesting phenomenon. Although the width of the particular diffraction line becomes sharp, the width of the diffraction line from the lattice plane perpendicular to it becomes wide. This is easily understood if we consider it on the basis of the Equation (4.1), indicating that the line width of Bragg reflection is inversely proportional to the average size of the crystallites in that direction. For instance, we consider the material consisted of the crystallite of board like shape of which surface is parallel to the (0 0 1) hexagonal lattice plane. In the direction perpendicular to the specimen surface, the Bragg reflections from the (0 0 1) lattice planes will be observed as having wide width, but those from the (h k 0) lattice plane are of sharp. If the external forms of crystallites are resemble one another and are oriented almost to the same direction, not only the change of line width but also anisotropy of intensity distribution along the diffraction rings occurs, depending on the hkl-index. This is due to the fact that crystallites of needle shape and those of plate like shape tend to orient to the same direction, as we can guess from the Figure 4.5 (b) and (c). As a natural consequence, powder diffraction rings obtained by those samples show remarkably anisotropic properties.

In order to observe the preferred orientation of the sample, the diffraction lines should be observed in a wide range of scattering angles. For this purpose, the powder diffraction camera with the X-ray film of plate or cylinder type is convenient to use. After checking carefully the existence of preferred orientation in the diffraction pattern observed, the information of the average size and the crystallite shape can be found by analyzing the widths of diffraction rings. Figure 4.8 is the 2-dimensional diffraction pattern of copper plate observed after a process of rolling of the sample material, where imaging plate has been used instead of usual X-ray film. The intensity is not uniform and modulating along each diffraction ring.

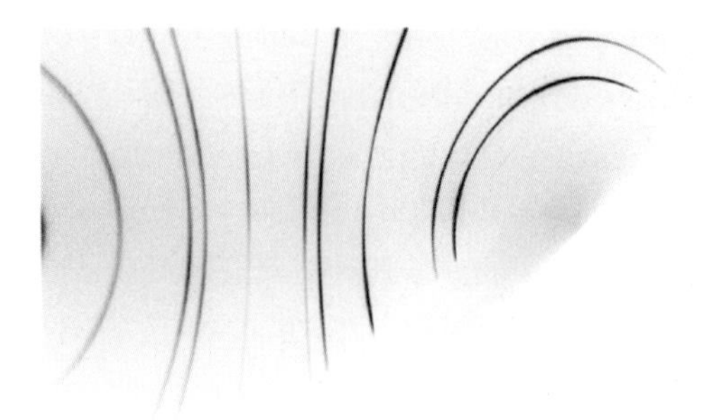

Figure 4.8 X-ray diffraction pattern of rolled copper plate.
Preferred orientation is observed.

However, the whole pattern seems to show some symmetry.

If utilized the diffractometer based on B-B type optics such as MiniFlex 300/600, could measure only the intensity distribution along an equator line of full diffraction pattern could be measured, as explained in Chapter 2. Thus, it may happen that some of the diffraction lines is observed of strong intensity, but some other diffraction lines are not. Such a matter can be easily readable from Figure 4.8. This is called as the diffraction pattern of preferred orientation. Recently correction formula have been introduced and have been utilized[9] to compensate the intensity distribution of the diffraction pattern of preferred orientation, which are called the **March-Dollase function** and the **spherical function**. This type of analysis became available by recent analytical software such as PDXL. Please refer to the reference.

Annotations

1) Crystallite size is often called particle size in textbooks. In the powder X-ray diffraction, a particle of the powder sample is very small but consists of a polycrystalline body, which is a collection of several of crystallites. This situation differs markedly from the crystallites in question. The "size distribution of crystallites", therfore, differes from "the distribution of grain size" which is often used in Japan. In order to determine the

distribution of grain size, X-ray small angle scattering can be used. On the other hand, the size distribution of crystallites can only be obtained by analyzing the profile of Bragg reflections. One should be careful about the difference of those two size distributions.

Part 2
Setting up of basic experimental tools

Part 1 presented several diffraction data measured by MiniFlex for well-known and discussed the data analysis. The goal of this discussion was to obtain detailed knowledge of the atomic-level structure of materials.

However, we have yet to describe X-ray sources, detectors, and optics, which are all important parts of X-ray diffraction instruments. Those topics will be addressed in Part 2. Additionally, the method of error evaluation for diffraction data is also included.

Chapter 5 X-ray source

Modern X-ray sources include synchrotrons, rotating anode sources, and X-ray tubes. The most often used X-ray source for "tabletop" experiments is X-ray tube. Before operating analytical instruments such as X-ray diffractometers, the operator should be familiar with X-ray sources and their principles of operation. Thus, this chapter discusses this instrumentation and the associated physics of X-ray diffraction.

5.1 How to generate X-rays

In the laboratory, the X-rays generated by X-ray tube are used for X-ray diffraction experiment. The principle of how to generate X-rays from the X-ray tube is shown in the Figure 5.1. This is a schematic view of the X-ray tube based on the Coolidge tube designed by W.D. Coolidge. The method of the X-ray generation has not changed since the discovery at that time. The inside of the X-ray tube has to be held as a vacuum. When the electric currents are flowing through the filament of left hand side, the filament is heated up and emits thermal electrons. The electrons are accelerated by a high voltage and collided to a **heavy metal target**. The X-rays are generated in the metal target collided by high speed electrons, and are emitted to all the directions. This state maintains as long as the filament is heated and a high voltage is supplied to the target. Since the X-rays generated are absorbed by wall of the glass tube, usually X-ray tube has four windows made of a thin beryllium plate to pass the X-rays for observations.

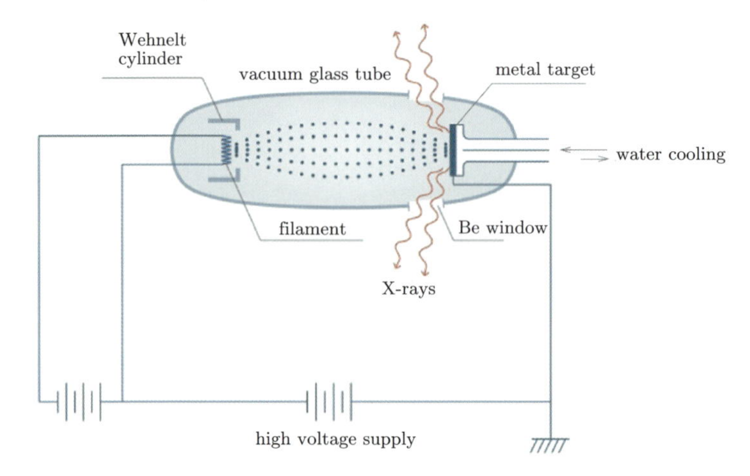

Figure 5.1 X-ray tube illustrating the principle of X-ray generation.

In the abovementioned process above, most of the kinetic energy of the electrons striking the target is converted into heat energy. A very small fraction of the energy is converted into X-rays. The X-ray generation efficiency is only 0.16 % of the supplied energy. Since most of the energy is converted into heat inside the target, the increase in the current causes the irradiated part of the target to melt, weakening the vacuum and resulting in electrical discharge. Both factors inhibit the stable generation of X-rays. Thus, the target should be water-cooled and be used within the range that the target will not melt. This is a critical power of an X-ray tube. The critical power depends on the target material and also the focus size of electron beam on the target. We thus compare X-ray tube performances by showing target metal types, X-ray sizes (focus sizes) and critical power.

5.2 Real nature of X-rays

X-rays were discovered by Wilhelm Roentgen in 1895, who named them so because they were mysterious rays that could penetrate through materi-

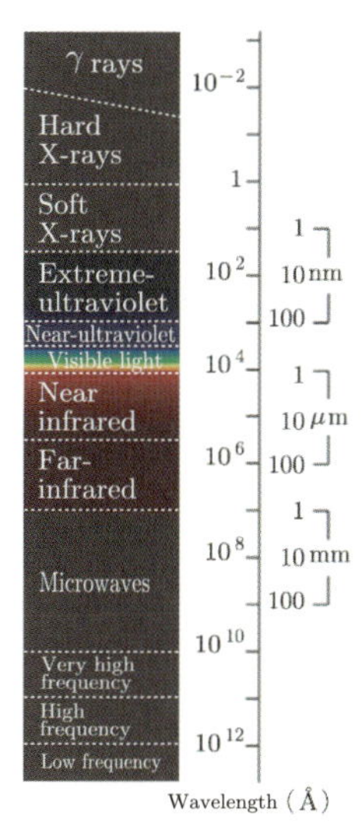

Figure 5.2 Classification of the electromagnetic wave.

als. The real nature of X-rays ,which is the same as that of electromagnetic waves as visible light, was understood by M. T. F. von Laue in 1912. However, the wavelengths of X-rays were known to be very much shorter than the visible light. The X-ray diffraction crystallography began with a pioneer study shown by W. H. Bragg and W. L. Bragg in the same year. The wavelengths of visible rays are several thousand angstroms (where an angstrom, or Å, is a standard international unit of measurement in X-ray diffraction crystallography; 1 Å is equal to 0.1 nanometer (1×10^{-10} m, 1×10^{-8} cm)). On the other hand, the wavelengths of X-rays are from 0.01 to 20 Å. The Figure 5.2 compares electromagnetic waves with various wavelengths. Because the X-rays are electromagnetic waves, X-rays show the same reflection and diffraction phenomena as visible light does.

X-rays behave like visible light, having characteristics of energy particles as well as electromagnetic waves. When you use a detector to detect X-rays, you will notice that the X-rays behave as particles. This is also applicable to visible light. In other words, detecting light will be an operation to confirm that light energy is received at some point, and that each light has reached there one by one. A light behaving as a particle with certain energy is called the **photon**. Then, in what state does the light of high intensity correspond? It can be said that the light of high intensity is in the state of the electromagnetic wave having high amplitude by the viewpoint of the wave nature. On the other hand, it can be indicated that many photons are concentrated in a certain position from the viewpoint of particle nature.

Let us imagine two kinds of diffraction or transmission images by X-ray films. One is an image obtained with usual exposure time using a known method. The other is the image merged by many images, each of which is taken for a very short time. Although each image measured for a short time is different one another, their merged image is in agreement with an image taken by a long time exposure. When an X-ray photon comes to a point on the film, it is observed as a simple dark dot there by receiving its photon energy. In the next moment, it comes to a different point and dark dot appears there. This process continues and so on. This is an experimental fact showing that X-rays are surely photons. Generally, the intensity of wave is given to be proportional to the square of amplitude. But, what do the intensity of X-rays represent? The answer is quite simple: it represents the probability of finding X-ray photons. A high X-ray intensity at a given point corresponds to a high probability of finding X-ray photons at that point. You should understand such a dual character of the wave and the particle of X-rays in this way.

With h as Planck's constant and ν as the frequency of X-ray electromagnetic waves, the energy of X-ray photons, ε, is given by $\varepsilon = h\nu = hc/\lambda$. This equation shows that X-rays are nothing but an aggregation of particles having this energy. Expressing wavelength λ in Å and energy ε in keV, we obtain

$$\varepsilon\,[\mathrm{keV}] = \frac{12.4}{\lambda\,[\text{Å}]}. \tag{5.1}$$

This shows that the energy of X-rays with a wavelength of 1 Å is 12.4 keV. The energy of the X-rays in medical use is relatively high and is more than 30 keV. As for the X-ray diffraction, the wavelength used ranges from 1 Å, which is almost the same order of magnitude with an atom, to 2 Å which is slightly longer than the atom: their energy is in a range of 7–20 keV.

5.3 Spectrum of X-rays

It is important to know the distribution pattern of the wavelength of X-rays generated by the X-ray tube and the distribution of intensity versus energy (called the spectrum). How we obtain this information will be discussed later. Here, Figure 5.3 (a) and (b) show the results obtained via numerical calculation by a researcher of RIGAKU. Figure 5.3 (a) shows the spectrums obtained for the three different targets: Cr, Mo, and W by fixing the supplied voltage to 33.5 kV. In Figure 5.3 (b), only the W (tungsten) target was used and the spectrums were obtained by varying the supplied voltage from 20, 25, 30, 35, 40, up to 80 kV. In Figure 5.3 (a), it shows a typical spectrum obtained by each target. The spectrum has two shapes; one is a gradually changing continuous spectrum in a certain region of wavelength and the other is a sharp spectrum at a particular wavelength. The former is called **continuous X-rays** or **white X-rays**, and X-rays with a sharp peak are called **characteristic X-rays**. Using the spectrum, we explain below a mechanism how these two different spectrums are generated.

When accelerated electrons hit a target of a metal plate, most electrons are consumed in heating up the metal plate and the kinetic energies of the electrons are eventually lost. Part of electrons change their orbits by

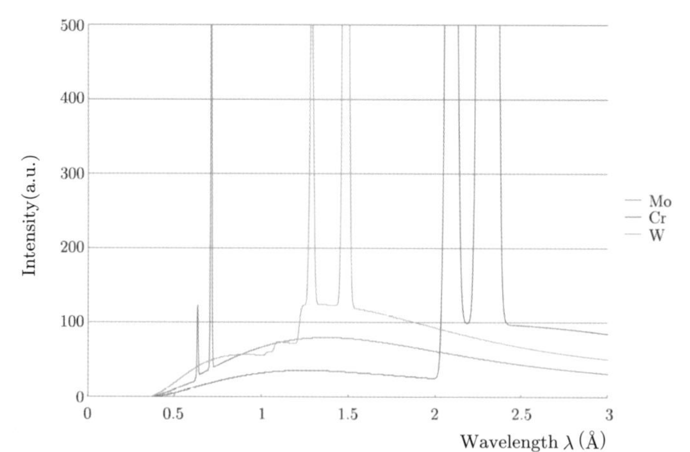

(a) X-ray spectrum from Cr, Mo, and W target by fixing accelerating voltage at 33.5kV

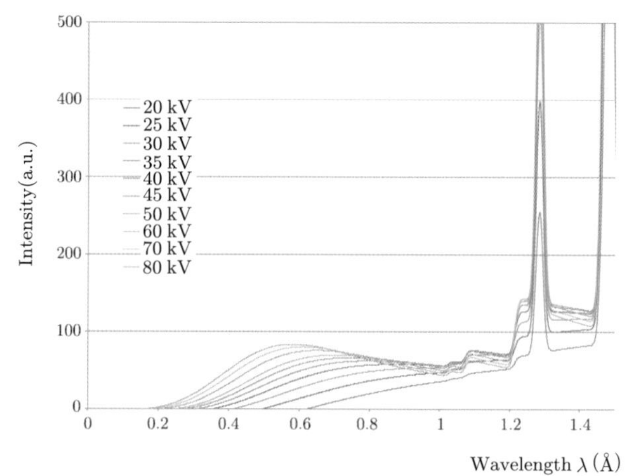

(b) X-ray spectrum from W target by changing accelerating voltage from 20kV to 80kV

Figure 5.3 X-ray spectrum.

strong electric field of atoms in a target. The X-rays are emitted along the tangential direction to the orbit. The electrons will lose their kinetic energy by the amount of energy used to generate X-rays and will decelerate.

This is called the **Bremsstrahlung** (German word: braking radiation). If there is surplus energy, the same process will be repeated and the X-rays will be emitted again. The X-rays which are generated from the process have various wavelengths and show the continuous spectrum, hence it is called as the continuous X-rays or white X-rays. The X-rays of the shortest wavelength corresponds to the case where the energy of incident electrons is all used as X-ray energy. It is called the **limiting wavelength**. Using the Equation (5.1), the wavelength is given by $\lambda_{\min}$ (Å) $= 12.4/V$ (kV). In the Figure 5.3 (a), as the applied voltage V is fixed to 33.5 kV, the shortest wavelength is 0.37 Å and it is independent of the kind of the target used.

The intensity of X-rays is in proportion to all the energy given to a target, namely, supplied power $W = IV$, where I is the current and V is the applied voltage. It is also proportional to the conversion efficiency ε ($\propto ZV$) and is obtained from IV^2Z. When the supplied power is fixed, the X-ray intensity increases with the increase of the atomic number Z of the target material. When you look at continuous X-ray spectrum shown in Figure 5.3 (b), you will notice the X-rays with long wavelengths particularly have low intensity. This is because X-rays of longer wavelengths are more easily absorbed until they reach to the surface of the target. This reduces X-ray intensity.

When accelerated electrons hit a target, their energies are used to produce X-rays by another mechanism. Accelerated electrons striking the target stimulate K-shell or L-shell electrons associated with the inner shells of metal atoms inside the target, prompting transitions and generating atoms that lack electrons (electron holes) in their inner shells. These atoms are ionized. When this happens, electrons in other shells within the same atom fall into the holes, emitting electromagnetic waves (photons) with energies in the X-ray region. The X-rays emitted from the target in this way are called the **characteristic X-rays**. Characteristic X-rays are described on the basis of the energy structure created by the electrons in the atoms. The

Figure 5.4 is a schematic view of energy level of an atom. When holes are produced in K-shell, X-rays emitted are called the K-series characteristic X-rays. Similarly, when holes are produced in L-shell, the X-rays emitted are called the L-series characteristic X-rays. The X-rays emitted in this way are classified into K, L, M, $\cdots$ series. There is a certain custom in order to express the characteristic X-rays in one-series. Subscript, α, β, γ are attached in turn from long wavelength to shorter wavelength. In an example, when the electrons of the L-shell and the M-shell fall down into the K-shell, the X-rays emitted are called as the Kα line and the Kβ line, respectively. Besides, the energy (or wavelength) of X-rays emitted is different, depending on the series to which the X-rays emitted belong.

Now, as seen in the Figure 5.4, L-shell and M-shell consists of multiple energy levels. Thus, the wavelength of the X-rays emitted as K-series is

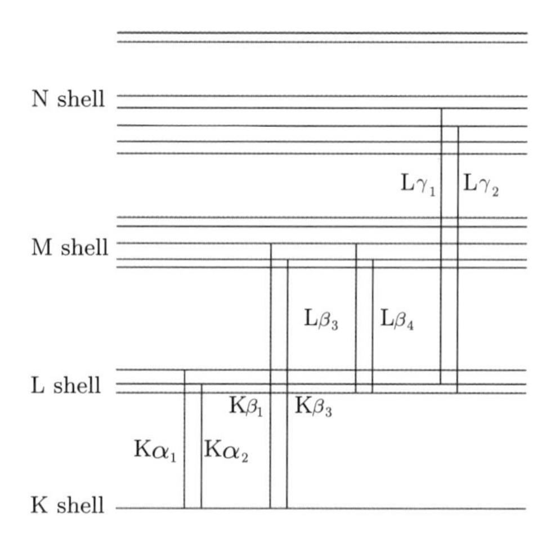

Figure 5.4 Generation mechanism of characteristic X-rays. The energy levels of electron within an atom and the characteristic X-rays expected based on the transition of electrons in upper level are drawn. Besides, there are other Kβ lines that are not drawn in this figure, because of negligible.

Table 5.1 Wavelength of characteristic X-rays Kα_1, Kα_2, $\langle$K$\alpha\rangle$, and Kβ. It is written by Kβ without distinction of Kβ_1 and Kβ_3, because they are observed as one spectrum with a broad width. These values refer to [10]. The unit is Å.

	V	Cr	Fe	Co	Cu	Mo
Kα_1	2.503610	2.289746	1.936081	1.789001	1.540593	0.709317
Kα_2	2.507430	2.293652	1.940019	1.792886	1.544414	0.713607
$\langle$K$\alpha\rangle$	2.504883	2.291048	1.937394	1.790291	1.541862	0.710741
Kβ	2.284446	2.084912	1.756645	1.620823	1.392246	0.632303

It is recommended to use the Cr/Fe/Co/Cu target for MiniFlex 300/600.

different, depending on the energy level: from which an energy level electron falls down to the K-shell. Additional subscript given as the numerical value is attached and distinguished in such as Kα_1 and Kα_2, where low numerical value shows that the X-ray intensity is strong: the intensity of Kα_1 is stronger than that of Kα_2. The intensity ratio $I(\mathrm{K}\alpha_2)/I(\mathrm{K}\alpha_1)$ is 1/2. The wavelength of Kα_1 is shorter than that of Kα_2. On the other hand, the intensity of Kβ_1 is stronger than that of Kβ_2, whereas the wavelength of Kβ_1 is longer than that of Kβ_2. The wavelengths of Kα_1, Kα_2 and Kβ are listed for six targets in the Table 5.1.

Among the characteristic X-rays, the intensity of Kα line is significantly strong so that it is used for the X-ray diffraction. When you begin testing the instrument, you would recognize that Kα line is actually made up with doublet consisting of Kα_1 and Kα_2.

5.4 X-ray tube voltage and current

When V is the voltage supplied to an X-ray tube and V_c is the voltage to excite the characteristic X-rays, the intensity of the characteristic X-rays I_c is proportional to the tube current i and to the n-th power of $(V - V_c)$.

$$I_c \propto i(V - V_c)^n \tag{5.2}$$

On the other hand, the intensity of the continuous X-rays, I_ω, is proportional to the current i, the square of the applied voltage V^2, and the atomic number Z.

$$I_w \propto iV^2Z \tag{5.3}$$

In these formulas, if the V is close to the excitation voltage V_c, the X-rays with a high intensity are not obtained. Thus, a larger voltage V is desirable. Empirically, it is recommended to take V two times or three times of V_c in the region of $n \sim 2$, and three times of V_c in the region of $n \sim 1$. In order to obtain a high intensity of the X-rays, the tube current needs to be increased under a tube voltage above a certain level. An appropriate tube voltage is shown in the Table 5.2. According to this table, the excitation voltage V_c of the Mo tube is high whereas that of the Cr tube is low. A high tube voltage for the Mo tube and a low tube voltage for the Cr tube should be therefore required.

What we consider is the intensity of the characteristic X-rays which are used in the X-ray diffraction. Although it is one idea to use a high applied voltage in order to obtain a high intensity, it becomes the cause of high intensity of unnecessary continuous X-rays. Because the intensity of the continuous X-rays is proportional to the square of the tube voltage and the tube current. If the tube voltage is significantly increased, the background in the X-ray diffraction data is increased; it is not good for obtaining high quality diffraction data. The tube voltage by which the intensity of the characteristic X-rays becomes maximum value is slightly different from that by which the ratio of peak intensity to the background (P/B) becomes large value. Please refer to the Table 5.2 which is the experimental result used by among the engineers at RIGAKU.

By setting the monochromator in front of the light-receiving side and the detector, the continuous X-rays can be removed. Thus, the P/B maximum condition in the table can be ignored since the P/B ratio is improved. It

Table 5.2 Optimal tube voltage.

Target	Excitation voltage (kV)	Appropriate tube voltage (kV)	
		Maximum X-rays intensity	Maximum P/B ratio
Mo	20	60	45 - 55
Cu	8.86	40 - 55	25 - 35
Co	7.71	35 - 50	25 - 35
Fe	7.10	35 - 45	25 - 35
Cr	5.98	30 - 40	20 - 30

is, however, recommended to use the P/B maximum condition in the case of the use of the Kβ filter. The details are described in the Chapter 7.

5.5 Absorption of X-rays

The X-rays penetrate well the material. It can be imagined from the radiograph. For the reason, the X-rays are used in various fields as well as the medical field as a light source of non-destructive inspection equipment. You may tend to consider that X-rays are not much absorbed by materials, although depending on materials. Actually, however, slight difference in materials will contrast X-ray absorption. You should realize that this advantage is rather utilized.

In the X-ray diffraction, the absorption should be considered in understanding the result of the quantitative analysis on the basis of the estimation of the depth of analyzing the sample. Thus, the absorption cannot be ignored. In addition, a high absorption material is needed for a radiation shielding of device which uses the X-rays. In anticipation of security, it is convenient if you know how to calculate the thickness to protect X-ray radiation. As a general property, X-rays with short wavelengths have a high penetrating capability. You should, therefore, pay careful attention to the leakage of X-rays. On the other hand, the X-rays with a long wavelength tend to be absorbed or scattered by the air. It means that it is easily absorbed by human body, although it may be safe if you are in some distance from X-rays. In the experiment, the absorption of X-rays by air is too large, not to be able to ignore if the wavelength of the X-rays is long. You may need to keep the device, light path or the like in a vacuum state.

So, let's consider the measurement of X-ray absorption. Let the decrease of X-ray intensity after passing through a material by the distance dx be -dI. This value is proportional to the incident X-ray intensity, I, giving the following equation:

$$-dI = \mu I dx \tag{5.4}$$

where μ is the proportion coefficient. Then, the Equation (5.4) is rewritten as

$$dI/I = -\mu dx \tag{5.4'}$$

In order to find out the intensity, the differential equation should be solved. If the intensity of the X-rays immediately before entering the material is I_0 and their intensity after traveling distance x in the material is I, we obtain the following equation:

$$I = I_0 \exp(-\mu x) \tag{5.5}$$

Here, μ is called the **linear absorption coefficient** and the unit is a reciprocal dimension of the length, 1/cm. The value μ/ρ, obtained by dividing the above value by the density ρ, is called the **mass absorption coefficient**. The values are provided in handbooks and other literature for all elements from hydrogen to lead against various X-rays with different wavelengths. The unit is expressed in cm^2/g. By referring to the values given, the mass absorption coefficient μ/ρ of chemical compounds and composite materials can be calculated using the following formula. Calculations can also be made for the linear absorption coefficient.

When the weight fraction of individual element is W_j, μ/ρ can be given by

$$\mu/\rho = \Sigma W_j \cdot (\mu/\rho)_{\mathrm{j}} \tag{5.6}$$

In a powder sample, the true absorption coefficient depends on the packing density in the sample container. It is generally estimated to be half the value of the corresponding solid material. It means that the absorption coefficient is multiplied by 1/2 in most cases.

For a given substance, the mass absorption coefficient μ/ρ changes depending on the wavelength of the X-rays. But, μ/ρ increases with the increase of the wavelength. This is shown in Figure 5.5. The horizontal axis indicates the wavelength of the X-rays. When the wavelength of the X-ray is a large value, that means it has low energy, the mass absorption coefficient is a large value, and the X-rays are hard to penetrate the material. On the contrary, the X-rays with a high energy are easy to penetrate the material. When the wavelength is fixed, the mass absorption coefficient of the materials of the heavy metal is larger than that of a light element.

In the Figure 5.5, there is the wavelength that absorption changes discontinuously. For instance, the Cu shows a discontinuous drop of the absorption at 1.378 Å: it corresponds to 8.998 keV. The wavelength of the discontinuous drop is shorter than that of Kα_1 line (1.541 Å) and Kβ line (1.392 Å). This wavelength with a discontinuous drop is called **the wavelength of K absorption edge** or **the energy of K absorption edge**. The absorption is increased with λ^3. The value drops to one digit smaller than that value at the absorption edge; the absorption increases again with a wavelength in proportion to λ^3. The reason for sharp drop at a certain wavelength is as follows. X-rays with short wavelengths are of high energy, much of which are used to excite fluorescent X-rays. Thus, the most of the X-rays are used for that purpose and are absorbed. On the other hand, X-rays of which wavelengths pass through the absorption edge do not have

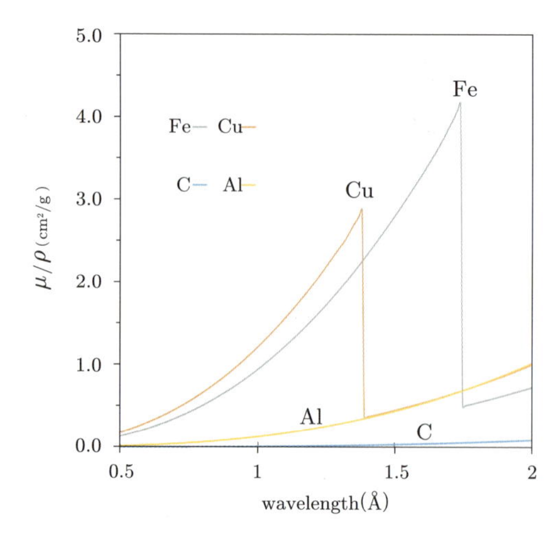

Figure 5.5 Wavelength dependent of mass absorption coefficient[10].

energy to excite fluorescent X-ray. This results in higher transmittance.

The absorption value depends on the atomic number Z and increases with the increase of Z in a form of Z^4. In addition, it happens that a part of X-rays changes the direction when entered into a material by the simple elastic scattering. In this case, the X-rays are detected as if absorbed. The amount of X-rays which are off from the direct beam are regarded to be absorbed. If this effect is included in, the mass absorption coefficient should be written by a sum of two terms: $(\mu/\rho) = C_1(Z^4/e^3) + C_2$. This part has not been included in the Figure 5.4. With the increase in the atomic number Z, the energy of the K absorption edge shifts to a higher energy (shorter wavelength). This is related to the increase in the energy of the X-ray fluorescence with the increase of Z.

5.6 Selection of X-ray tube

The Figure 5.6 shows the (1 1 0) diffraction peaks of α-Fe powder sample measured by five X-ray tubes. All data were obtained by the use of the Kβ filter method which is mentioned in the Section 7.6.2. The wavelength of the Kα line of the Mo tube is the shortest, and the wavelength gets longer in order of Cu, Co, Fe, and Cr tube. Because of it, the diffraction angle moved to high angle. The reason why Cu tube becomes the standard in powder X-ray diffraction is that high intensity diffraction pattern can be obtained and this wavelength is relatively easy to use. When the wavelength is too short, the reflection peaks are densified. On the contrary, the number of the observed peak is limited if the wavelength is too long. From the figure, we can find that the peak intensity and the P/B ratio depend on the kind of the tube. The main cause of different background value is due to the fluorescence X-rays emitted from the sample.

Therefore, a sample including Fe will create relatively weak diffraction peaks when using Cu tube, because the X-rays from Cu tube are exhausted to exite easily the Kα X-ray fluorescence of Fe. It is cause for a high

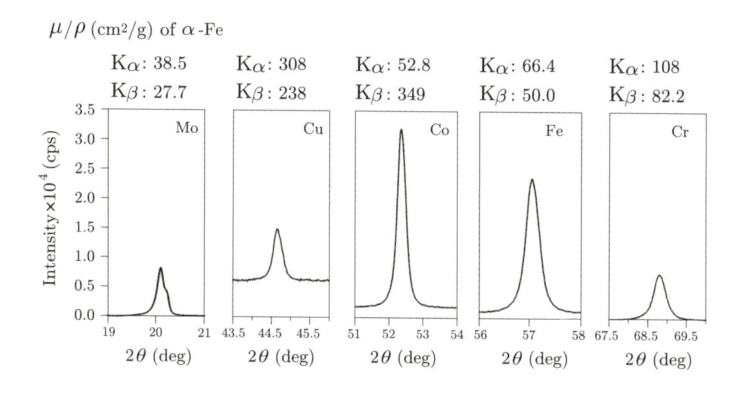

Figure 5.6 Change of the diffraction angle and the P/B ratio by the target material of X-ray tube.

background and a low intensity. Therefore, for a sample including Fe, it is recommended to use Fe or Co tube with the use of Kβ filter.

In the measurement with the use of monochromator at the detector side, it is possible to remove not only the Kβ but also the fluorescence X-rays. Even if any sample including Fe is measured by using Cu tube, the data of low background can be obtained. However, as shown in Figure 5.6, the Co tube is advantageous to use in the case of Fe-based materials, because its diffraction lines are several times stronger than that of Cu tube with monochromator at the detector side.

Chapter 6 X-ray detector

Detection of X-rays is performed by changing X-rays into a detectable form using the interactions between X-rays and meterials. After glancing a variety of detector that has been variously used until now, scintillation detector and also the measuring system will be described, in conjunction with D/Tex Ultra, which has been recently invented as a one-dimensional semiconductor detector.

6.1 X-ray detectors

Various X-ray detectors currently in use are listed below:

(1) Photographic detectors: Use of photograph films and dry plates. They are rarely used.

(2) Ionization detectors: Use of ion chamber, Geiger–Müller counter, proportional counter, and position-sensitive proportional counter.

(3) Photoluminescence detectors: Use of a fluorescent screen, scintillation counter, charge-coupled device, and image plate.

(4) Semiconductor detectors: Silicon drift detector (zero-dimensional), semiconductor-strip detector (these are one-dimensional devices, D/teX Ultra), and high-resolution, high-speed two-dimensional photon-counting X-ray detector (e.g., HyPix-3000). Use of electron transitions between the valence and conduction bands and include solid-state detectors.

Another detector that has been in use is the Image Orthicon, which uses photoconductive effect and X-ray television. These are not discussed in detail.

6.2 X-ray film

Because X-rays expose photographic emulsion, they may be detected using photography methods. In this method, the detecting position and X-ray intensity and location are recorded on the film, which may be conveniently stored in a notebook. However, the sensitivity and the digitization of X-ray films are inferior to those of the counters.

The photograph method is now used rarely, because the processing required to develop the film is laborious and has raised non-negligible environmental issues. Today, such photographic film is being replaced with photostimulable phosphor film to form image-plate detector, which combine improved reliability and digitization.

6.3 Geiger–Müller counter

The operating principle behind a **Geiger–Müller counter** is ionization of a gas. These detectors normally comprise a glass tube filled with gas. At the center of the tube is a wire, which serves as an electrode. This electrode is maintained at a voltage of 100 V with respect to the gas. When X-rays interact with the gas, atoms or molecules in the gas have a certain probability to be ionized, which generates electrons and ions. The electric field generated by the electrode drives electrons toward the wire, where they are collected and detected as a signal whose amplitude is proportional to the quantity of incident radiation (regardless of its energy, provided it is sufficient to ionize the gas). This type of counter is called an "integrating detector" and is used to detect the intensity of radioactivity as well as X-rays.

When operated at a low voltage, a Geiger–Müller counter produces a signal that is proportional to the energy of the incident radiation; in this case, this is called **proportional counter**. This is used still conveniently used to detect and analyze X-ray fluorescence. By processing the signal generated in the core wire from both ends of the core wire, the position on the wire at which the signal originated may be deduced. This approach results in a **position-sensitive proportional counter**. Such detectors may be made two-dimensional by fixing the anode wires in a parallel arrangement at a given interval.

6.4 Fluorescent screen and image plate detector

Some materials (e.g., ZnS) emit light after absorbing the energy from radiations including X-rays. To absorb the energy of the incident X-rays, ground-state electrons in the material transfer to excited states; when they return to their original state, they emit light (normally of a longer wavelength than the incident light).

Several types of instruments exist to detect fluorescence: fluorescent screen, scintillation counter, and image plate. The scintillation counter is described in Section 6.5, whereas the fluorescent screen and the IP are briefly presented here. The fluorescent screen consists of a cardboard caked with a binder and ZnS(Ag) powder (Ag is the activator). NaI(Tl) is also often used as fluorescent substance for detecting X-rays. The fluorescent screen was used to detect from which direction X-rays appeared. However, such detectors are rarely used anymore.

The image plate also exploits fluorescence but in a slightly different way. A film is coated with a layer of photostimulable phosphor [BaFX: Eu^{2+} (X = Cl, Br, I)]. When irradiated by X-rays, the irradiated location stores the X-ray energy by producing what is called a latent image. When the film is again irradiated by a laser at a certain wavelength, the stored energy is released in the form of light (i.e., the latent image is released). This photo- stimulated fluorescence is proportional to the energy deposited by

X-rays at each given point in the irradiated film. Thus, the emitted photostimulated fluorescence forms a two-dimensional pattern that reflects the spatial distribution and the intensity of the original X-ray radiation. By scanning the latent image with a laser at the an appropriate wavelength, this image may be digitized, which facilitates storage and analysis. Because the diameter of the laser beam determines the spatial resolution of the image, the resolution of the image-plate technique is less than that of X-ray film. However, image-plate sensitivity is high, and its dynamic range is very large due to a saturation count that reaches seven digits, which are generally considered superior to X-ray film. Image plates are widely used for medical and life-sciences applications.

6.5 Scintillation counter

Scintillation counter (see Figure 6.1) also exploits fluorescence from solid materials. A block of such material, called a "**scintillator**" is placed in front of the photocathode in **photomultiplier** tube. When X-rays photons deposit their energies in the scintillator, it emits light at a longer wavelength. Some of these photoluminescence photons strike the **photocathode** and create photoelectrons via the photoelectric effect. These photoelectrons are accelerated toward another electrode, which they strike to release yet more electrons. A succession of such electrodes allows this process to repeat, thereby multiplying the number of released electrons in a geometric progression. A single photoelectron can thus be multiplied by six orders of magnitude to create a final output pulse of several mV.

A common scintillator material is NaI(Tl), which consists of single-crystal NaI with trace amounts of Tl. The crystal has strong deliquescence characteristics, is difficult to handle. It emits in the violet. The photoluminescence emitted by a NaI(Tl) scintillator is proportional to the energy of the incident X-ray photon, which means that the peak voltage of the electric

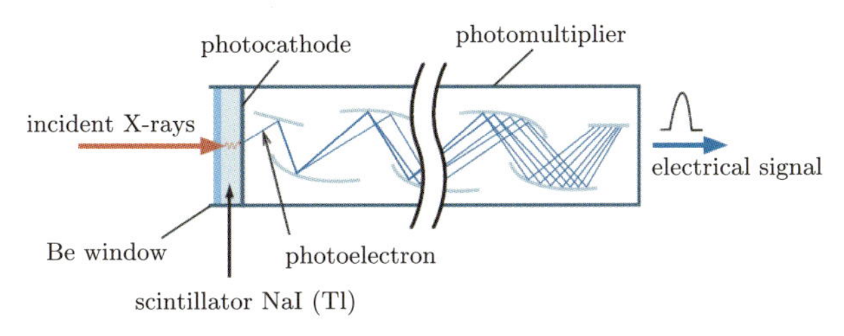

Figure 6.1 Scintillation counter.

pulse from the associated photomultiplier tube is proportional to the incident X-ray energy. This makes such tubes useful for analyzing the energy of X-rays.

A disadvantage of scintillation is its relatively high signal-to-noise ratio. For X-rays with wavelength $\lambda \geq 3$ Å, the signal-to-noise ratio can approach unity, making it difficult to separate signal from noise. Fortunately, this problem seldom arises in practice because the wavelength of most X-rays used to analyze materials is less than that of the CrKα line (~2.29 Å).

The count of a scintillation counter is calculated by dividing the number of converted pulses (not peak pulse voltage) by the number of the incident X-ray photons. The efficiency is close to 100 % in the wavelength range for X-ray diffraction (−1 Å). Most X-ray photons that enter the scintillator crystal cause photoluminescence and are thus converted into an electrical signal. The graph of count as a function of incident X-ray wavelength (Figure 6.2) shows that count increases with wavelength. However, when the wavelength exceeds a certain threshold, the count worsens because X-rays are absorbed by materials in the light path or windows. For comparison, Figure 6.2 graphs count vs incident X-ray energy for both a Geiger–Müller counter and a proportional counter.

The energy resolution of a counter depends on its intrinsic characteristics. For instance, a proportional counter can be used to count monochromated

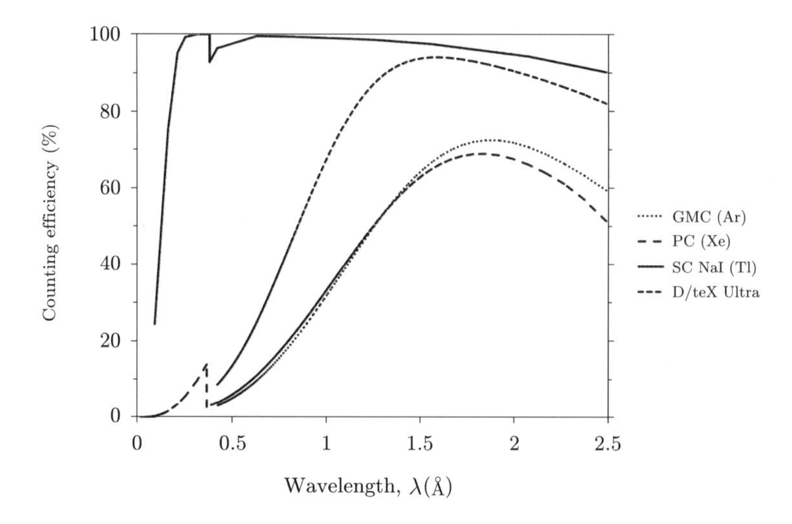

Figure 6.2 Efficiency of detectors. The efficiency of the scintillation counter is significantly higher than that of the gas proportional counter. The figure shows a difference between gas and solid detector element.

X-rays with an extremely narrow energy distribution with a 20 % pulse width. The energy resolution of scintillation counter depends on the statistic of the process by which X-ray photons are converted to an electrical signal. The estimated energy resolution of a scintillation counter is not very good. In this case, the problem comes from fewer primary electrons being generated at the photocathode. At best, only 10–15 electrons are generated when a single CuKα is used.

The statistical fluctuation in signal strength is given by $\sqrt{N} = 3 - 4$, where N is the number of photoelectrons ejected from the photocathode. Calculation of $\sqrt{N}/N \times 100\,\%$ gives $1/\sqrt{N} \times 100\,\% \approx 25 - 30\,\%$. The total energy resolution is 40–60 % after taking into account the noise outbreak in the latter process.

6.6 Semiconductor detector

The electronic structure of semiconductors is dominated by the valence band, which is almost fully occupied by electrons, and the conduction band, which is almost empty. The "bandgap" is the energy difference between the lowest-energy state of the conduction band and the highest-energy state of the valence band. Although the bandgap varies between semiconductors, it is typically several of the order of eV and is always very much smaller than the typical ionization energy of atoms. When radiation is absorbed by a semiconductor, electrons from the valence band transfer into the conduction band, which is analogous to an atom being ionized. As a result, positive holes (where "hole" refers to the absence of an electron) are added to the valence band, and electrons are added to the conduction band. The formation of an electron–hole pair may be thought of as the "ionization" of atoms in the semiconductor (although the bulk semiconductor material remains electrically neutral because the freed electrons do not escape from the material). If the semiconductor material is used as an electrode in a proportional counter, the holes in the valence band and the electrons in the conduction band accumulate and produce a signal that is used to detect the "ionizing" radiation. This is the basic physical mechanism behind semiconductor photodetectors.

Because the bandgap in semiconductors is quite small (-1 eV), electron–hole pairs created by the incident X-rays is a large number to which is the height of the output electrical pulse is proportional. The incident radiation energy can thus be analyzed on the basis of the height of the output electrical pulse. Characteristics of this semiconductor detector are their high-energy resolution and high sensitivity, compared with other detectors. The only disadvantage is that thermal noise is easily picked up due to the small bandgap energy. So this detector was used by cooling its detecting element with liquid nitrogen.

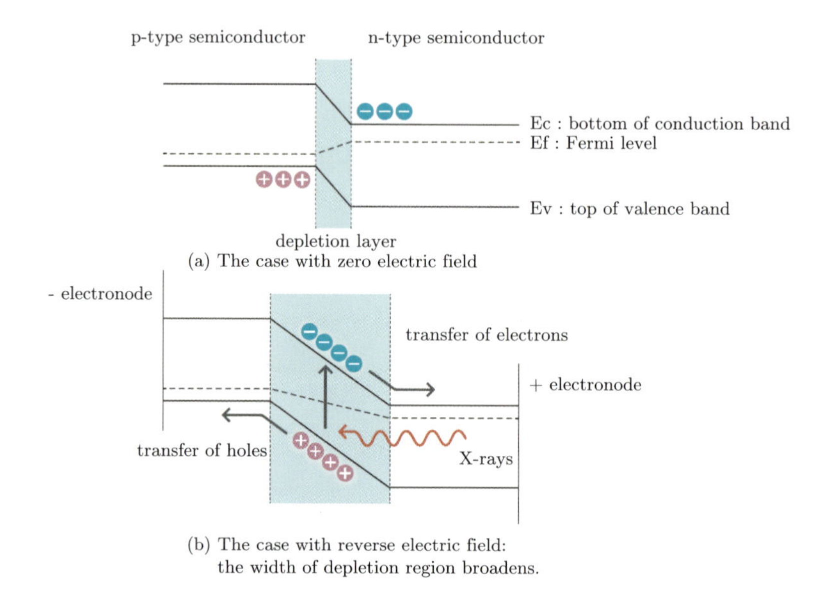

Figure 6.3 Electronic state in the depletion region of the p-n junction photodiode.

The solid state detector SSD based on Si(Li) and Ge were introduced for use in X-ray diffraction in the early 1970s. Their high-energy resolution (about 150 eV) was exploited to study Compton scattering and diffraction over a broad spectral range by using bremsstrahlung X-rays. In other work, a silicon drift detectors SDD were examined to analyze fluorescent X-rays. In the 1990s, such detectors were downsized, and the energy resolution was improved; this is notable due to the development of the PIN photodiode and the avalanche photodiode. Since then, detecting elements have become remarkably integrated with analytical instruments, creating today's **multidimensional detector**: Some such as instruments include PILATUS (developed by DECTRIS), HyPix-3000 and D/teX Ultra (by RIGAKU) which is used in the MiniFlex 300/600.

Now, the operating principle of p–n junction photodiodes will be explained. Figure 6.3 (a) shows the energy diagram at the p–n junction with no external electric field. A depletion layer forms at the p–n junction. Applying a reverse bias across the p–n junction [i.e., lowering the voltage on the p^{++} side and increasing the voltage on the n^{--} side; see Figure 6.3 (b)] causes more holes (electron) to leave the p (n) side, thereby widening the depletion layer. When an X-ray photon is incident on the depletion layer, it liberates a large number of electrons and holes, with the total being proportional to the energy of the original X-ray photon (the large number of electron–hole pairs created for a single incident X-ray photon leads to the excellent energy resolution of p–n junction X-ray photodetectors). These newly created electrons and holes are separated by the electric field within the depletion layer, as seen in Figure 6.3 (b), thereby creating an electric-current pulse between electrodes of the element, which this signals the photon-absorption event. This is the basic operating mechanism for semiconductor detectors made from p–n junction photodiodes.

The number N of electron–hole pairs created by an incident X-ray photon is defined by the incident X-ray energy E_0 divided by the average formation energy ε for electron–hole pairs: $N = E_0/\varepsilon$. In this equation, ε is about three times the bandwidth (3.64 eV for Si, and 2.96 eV for Ge). This value is an order of magnitude less than the typical ionization energy of gas. For the interested reader, more details are available in standard textbooks.

Similarly to scintillation counters, the semiconductor detecting element absorbs the radiation, which leads to large counting rates. However, for Si- and Ge-based devices, short-wavelength radiation may penetrate the detector material without being absorbed, so the efficiency becomes low. This effect is reflected in the catalog of RIGAKU showing that, for the D/teX Ultra, the detection efficiency is about 99 % for Kα radiation from Cr, Fe, Co, and Cu and is approximately 35 % for Mo. It shows that the energy resolution for the CuKα line (wavelength $\lambda = 1.542$ Å, energy $h\nu = 8.04$ keV) is 1.6 keV, which is almost 20 %. This is the same level of precision as that of the proportional counter and very much better than the scintillation counter of which resolution is in 35–40 %.

6.7 Counting circuit

Although the process of transforming X-ray photons into electrical signals differs for proportional counters and scintillator counters, the output of both devices is similar. For each device, the peak of the voltage output (called pulse height) is proportional to the energy of the X-ray photon detected. Furthermore, the number of X-ray photons of given energy detected by the device is obtained simply by counting the electric pulses with the corresponding pulse height. Figure 6.4 shows a block diagram of the electric circuit used to count X-ray photons.

In any case a **preamplifier** is integrated into the detector close to where the electrical signal is created. In Figure 6.4, the preamplifier and the detector are bound by a single block. The signal from the detector goes to the **linear amplifier**, which is placed before the counter to reduce noise. After amplification by the linear amplifier, the signal is transferred to the counter.

After having its waveform straightened by the linear amplifier (which

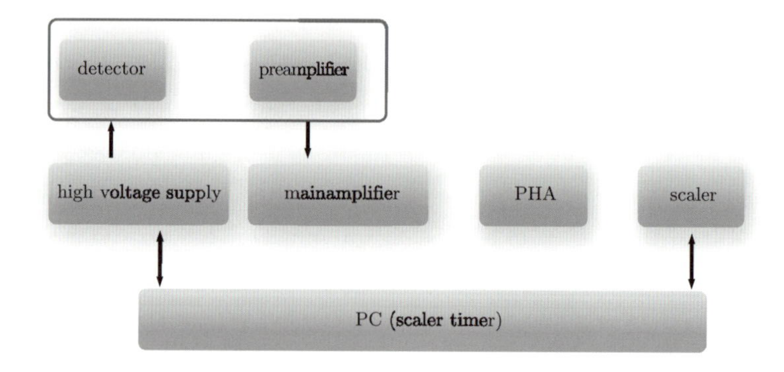

Figure 6.4 Block diagram of counting circuit.

determines the temporal width of the wave pattern), the signal is amplified and transmitted to a **pulse-height analyzer** (PHA). After discriminating against pulses of undesired height, the signal is sent to the counter (i.e., **scaler** in Japan), which counts the number of pulses that occur over a given time.

We now explain the function of PHA in detail. X-rays scattered from a sample are detected by a scintillation counter, and the height of the resulting electrical pulses are analyzed by the PHA. The illustration of the distribution of pulse heights treated by the PHA appears in Figure 6.5, which shows the number of incident pulses as a function of pulse height. The curve is called the **pulse–height distribution curve**, which is proportional to the energy spectrum of the incident X-rays.

The range of the pulse–height distribution curve can by modified by tun-

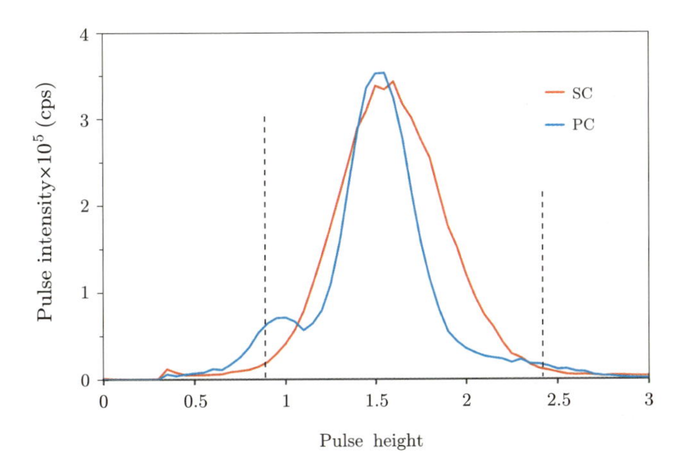

Figure 6.5 Pulse height distribution curve obtained through the PHA, after detecting CuKα radiations by SC and also PC. Red and blue line represents the results by SC and PC, respectively. The horizontal axis is the pulse height of incident X-rays, and the longitudinal axis is the number of pulse detected.

ing the gain of the amplifier. This can be done by changing the applied voltage of the detector, or by changing the scale (scale of the horizontal axis). Figure 6.5 shows how varying these parameters affects the pulse–height distribution curve. As mentioned previously, the pulse–height distribution reflects the X-ray-energy distribution. Thus, noise from low- and high-energy X-rays can be removed by only using pulses within the peak of the pulse–height distribution.

In general, the PHA allows the pulse–height threshold to be set; pulses with pulse height below (above) the given height are rejected (accepted), which is called **discrimination**. By using this function, one may discriminate against pulses with pulse height below the dotted line that appears on the left side of the figure and against pulses with pulse height above the dotted line that appears on the right side of the figure. A PHA with such a double-discrimination function is called a "multichannel analyzer" (MCA). Next, the pulse width threshold is set, which is called the **window width**. By proper adjustment of the window width, only the pulse signal will be passed to the scaler. It must be noted that there may happen such a case in which the energy spectrum of incident X-rays is not faithfully reproduced by the PHA.

Finally, electronic noise is normally present around zero pulse height; however, this noise is removed by a purpose-built electronic circuit.

6.8 Two-dimensional pixel-array detector

Because semiconductor detectors are very sensitive, the detection element can be miniaturized. Nowadays, the X-ray sensitive area can be reduced down to 100 micron meter square while maintaining the counts per second (cps) above 10^6 cps. In other words, X-rays that deposit energy into the micron-sized X-ray sensitive part of the detector can be measured at the rate of 10^6 cps, even when the active semiconductor is $0.1 \times 0.1\,\text{mm}^2$. X-ray-detection elements are so small that they can be integrated onto Si substrates as one-dimensional or two-dimensional arrays. This exemplifies another characteristic of semiconductor detectors: They can be integrated in this way or even used in integrated circuits.

Note that further reducing the size of the active detection element would not necessarily be welcomed because the size of the counting electronics and the aforementioned scaler would also need to be reduced. Fortunately, the integrated counting circuit may also be further miniaturized (see, for example, the counting block encircled by the broken line in Figure 6.4), and the counting circuit and detector can be integrated into the form of one-dimensional or two-dimensional arrays. Combining these two integrated boards leads to a composite-type multidimensional detector, which is outlined in Figure 6.6. This schematic drawing shows a measurement system array, including power supply, in the lower part, which is connected to the upper part by bump bonding. Such detectors are called hybrid **pixel-array detectors**.

Figure 6.6 shows a section of the semiconductor pixel detector. When the

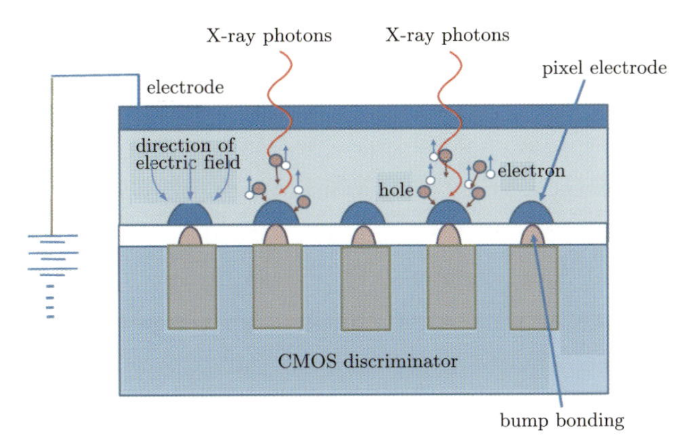

Figure 6.6 Schematic view of cross section of the hybrid two dimensional pixel detector.

reverse bias is applied to p^{++} n^{--} junction, the depletion layer is slightly expanded. When the X-rays are absorbed in such layer, a lot of electron-hole pairs are created. Within the two carriers, the holes move along an electric field and reach to an electrode which is in a dot shape and forms a pixel. An electric signal in the pulse shape is generated there and is transferred to the integrated counter element of the lower part through the bump bonding.

6.9 One-dimensional semiconductor detector

The D/teX Ultra X-ray detector is a one-dimensional semiconductor detector that can be used with MiniFlex 300/600. Its sensitive area is made as narrow as possible in the horizontal dimension. Because a high spatial resolution is required, each detector element is 0.1 mm wide and 20 mm long; this gives the overall active detection area in a strip form. Thus, this single-strip detector constitutes a single channel, and the entire detector is made of 128 such single-strip detectors (for 128 "channels"). The total length of the active detection area is thus 12.8 mm.

As illustrated in Figure 6.7, each channel is wire bonded to linearly arrayed counters, and the accepted X-ray energy range can be set by adjusting the pulse–height discrimination, baseline, and window width. The X-ray energy resolution is less than 25 %, which compares well with that available from a scintillation counter, which is 40 – 60 % and is as good as that available from proportional counters. Thus, measurements can be made even of materials such as iron by suppressing the background of fluorescence X-rays (assuming a CuKα X-ray source is used).

The counter efficiency of the D/teX Ultra exceeds 95 % for CuKα X-rays. This detector can detect CuKα X-rays even if the intensity ratio of the diffraction line is several digits. Although this chapter focused on detector

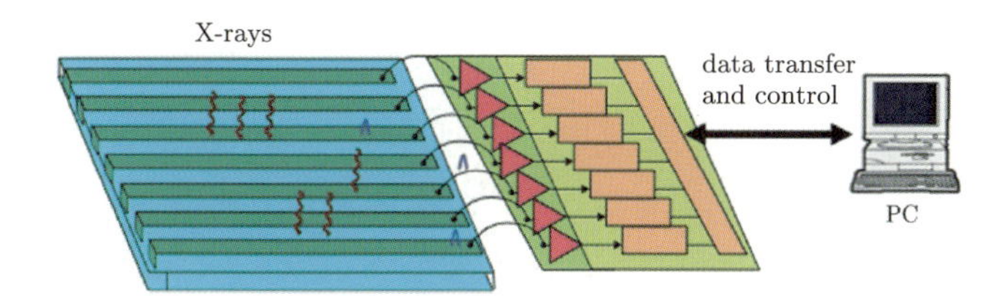

Figure 6.7 Schematic view of one dimensional detector.

issues, Chapter 7 discusses in detail the optical systems used with detectors for powder diffractometry and their various advantages and disadvantages.

Chapter 7 X-ray optics

Ideally, the X-ray diffractometer should operate with parallel beam optics. To approach these conditions, we must reduce the sizes of the light source, the sample, and the slits. Most of the X-rays emitted from the X-ray tube will then be omitted. Among the optics currently available, the focusing optics used in the convergent beam method is most effective. The more we understand these optics, the more we appreciate their superior performance. Nowadays, the resolution can be reproduced by a computer with submitting of fundamental parameters to the available software. In this way, the precision can be customized to the purpose, and the device can be managed. This chapter explains the optical elements required for a useful convergent beam method.

7.1 Slits and their role

When observing diffraction data by a two-dimensional detector such as a film or a counter, the incident X-rays should be parallel with a small cross-section. The parallel beam is realized by a component called the **collimator**. The parallelism of the X-rays diverged from a **point focus source** is adjusted by a pinhole or an up–down–left–right slit placed at distance L from the source. This slit is called a **divergent slit (DS)**. Given a X-ray source of size ΔX and a slit size of ΔS, the divergence angle $\Delta\theta$ of the X-rays can be calculated from the geometry and is approximated as $(\Delta X + \Delta S)/2L$. The angular divergence $\Delta\theta$, which indicates the resolution of the detector, can be improved by reducing both ΔX and ΔS; however, this action also reduces the intensity of the parallel beam, which is inversely proportional to the square of L.

To overcome this disadvantage, the point source is replaced by a **line focus source**. The line source provides a large divergence angle in the vertical direction of the X-rays, but reduces the resolution in that direction. Therefore, instead of a simple thin slit, multiple slits are piled up at equal intervals in a thin plate-formed shield. This is equivalent to arranging several point focus sources in the vertical direction and prevents vertical divergence. In this configuration, called **Vertical Soller Slits (VSS)**, the intensity of the X-rays increases proportionally to the number

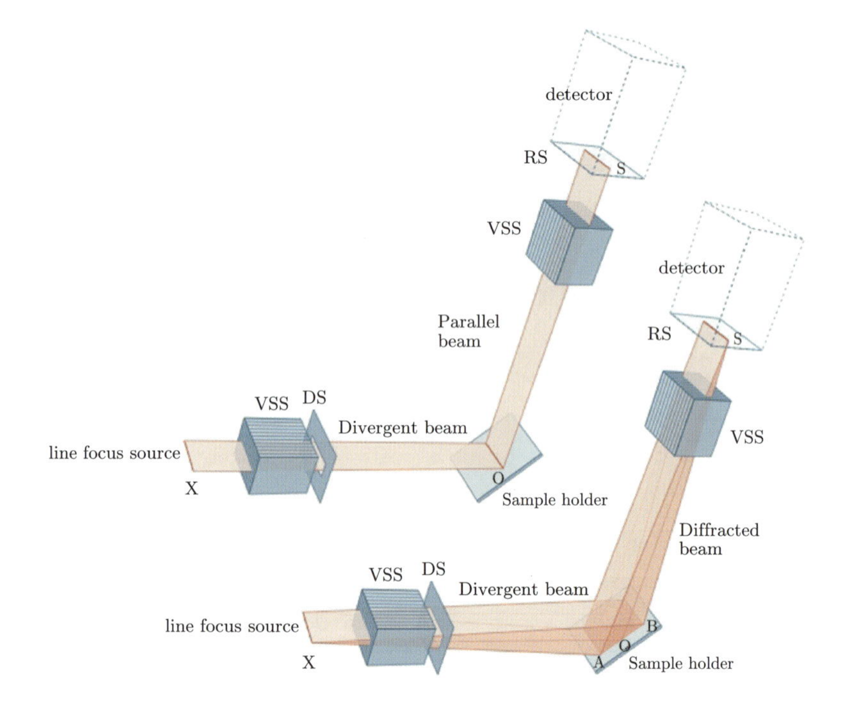

Figure 7.1 Schematic view of VSS.

of slits. Note that the "vertical" direction in VSS is the direction orthogonal to the diffraction plane; that is, the flat surface formed by the incident and diffracted X-rays. The top panel of Figure 7.1 is a schematic of the VSS optics, showing the basic principles. The Soller slits are arranged in the vertical direction. The diffration can be measured form line shape region (instead of are point) on the sample. Such meacurement is enable by the Soller slits as shown in Figure 7.1. As shown in the bottom panel of Figure 7.1 and explained in the next section, the diffracted X-rays can be collected from the entire surface of the plate sample. The measurement efficiency can be increased by adding a counter.

The **receiving slit (RS)** is placed in front of the detector. The VSS is then placed in front of the RS to remove any extra vertical scattering from the sample. Although not drawn on the figure, the quality of the diffraction line can be improved by placing additional slit Pr (see the Panel(a) of Figure 1.1) that prevent X-rays scattered from outside the measurement area of the sample.

7.2 Convergente X-ray beam optics

The bottom panel of Figure 7.1 shows the X-ray optics for convergent beam methed. It is often called as the **Bragg–Brentano optical system**, along with the convergent system. Here after B–B optics will be used as an abbreviation. Figure 7.2 shows the projection of these optics onto a diffraction surface. The incident X-rays (XO) and the scattered X-rays (OS) are symmetric at the sample surface AB with a plate type; the system is designed to allow variable scattering angle 2θ. Therefore, as the sample stage AB is rotated in the θ direction, the optical system including the detector (often called the counter arm) rotate through 2θ. The distance from the symmetric condition is given by $|XO| = |OS|$. The X-rays that have departed from the X to the O direction produce a Bragg reflection at

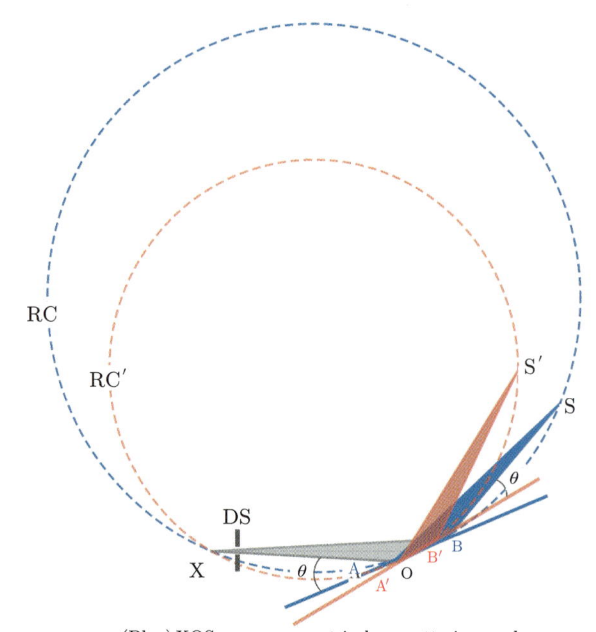

(Blue) XOS : arrangement in low scattering angle

(Red) XOS' : arrangement in high scattering angle

Figure 7.2 Optics of the convergent beam method, called also the B-B method.

the center position O of the sample, and the diffracted X-rays turn toward a slit S.

The three points X, O, and S lie on one circle as shown in Figure 7.2. This circle is called the **Rowland Circle** (labeled RC in the figure). When A and B are on the circle, the X-rays directed from X to A (or B) are scattered in the 2θ direction. By plane geometry, we can understand that the circumferential angle subtending arc AB of the circle RC is constant. Thus when the beam size widens, all the diffraction lines from the crystallites on arc AB of the sample are focused on the point S. In addition to the optics, the exposed surface of a sample may be two-dimensionally extended by combining the line focus X-ray source with VSSs. This system is very useful for obtaining high-intensity data. A powder sample should thus be fixed along an arc.

Let us consider changing the diffraction angle. The diffraction direction OS′ makes a high 2θ angle with the incident direction XO, as seen in Figure 7.2. The sample position O is equidistant between the X-ray source X and the receiving slit S. Both distances are called the **camera length**. As the camera length is constant, changing the diffraction angle means changing the radius of the Rowland circle. Although changing the diffraction angle alters the arc AB to the arc A′B′, the difference is expected to be small, so we reploce the arc AOB by a plate. Vagueness is problematic because the diffraction line does not strictly forcus on S. It should be understood that the optical system of B–B method is based on above approximations. In addition, one Soller slit VSS and a divergent slit DS are inserted in the arm of the X-ray incident side of the device, another is positioned the VSS in the arm of the X-ray receiving side is symmetrically located before the focusing point RS. Thus, detector should admit X-rays over vertical divection of about ten millimeters. Bragg–Blentano type. X-ray optics converge the X-rays diffracted from the wide area of sample just before the counter, and hence, constitute a **convergent beam method**. The MiniFlex 300/600 diffractometer is of this type.

7.3 Aberration and resolution

Compared with parallel beam method, the B–B method has the following merits and demerits. This method increases the intensity of the diffracted X-rays by more than one order of magnitude and ensures a highly efficient diffractometer. However, when querying the diffraction profile or precisely measuring a diffraction angle by B–B method, much care is required.

When the sample surface describes an arc along a focusing circle (the Rowland circle), as explained in the previous section, the X-rays diffracted from one point on the focusing circle collect at position RS of the receiving slit symmetrically around O, the central position of the sample. Because the radius of the focusing circle changes with the angle θ, the radius of the arc described by the sample surface should change accordingly. However, the arc of the sample surface is difficult to change in practice, so the B–B method adopts the plate-type configuration, not the arc shape. Therefore, once the X-rays diverged from one point on the X-rays source have diffracted from the sample surface, they are not always collected at one point on the receiving slit. The collection position depends on the diffraction position on the sample surface and is dispersed around the RS point. This problem with B–B is inherent and cannot be avoided. Additionally, the diffraction angle is sensitive to location changes on the sample surface. The resulting vagueness is analogous to that of optical lenses and is given the same name, **aberration**.

The intensity distribution of the diffraction line measured by a θ–2θ scan. The distribution line has a certain width (originally much narrow) around a given value of 2θ. The intensity distribution, which is a function of scattering angle 2θ and makes the width of the diffraction line spread, is called the **resolution function**. The resolution is peculiar to the equipment, so it changes with the width of the X-ray source, divergence angle of the incident X-rays, and slit width at the acceptance side. Therefore, in order to see the resolution function, the intensity distribution should be measured on a standard sample with a large particle size. In stress or strain measurements, which need precise observations of the lattice spacing, through the ages parallel beam method is recommended because even slight shifts in the diffraction angle cause serious problems. Although aberration is not limited to the B–B method, it requires special attention when applying this method.

Another problem deserving attention occurs in diffractometers using a line-focus X-ray source, as shown in Figure 7.1. The VSS removes the divergent influence of the X-rays in the vertical direction. At sufficiently short distances of the Soller slit, the vertically arranged beams become essentially parallel, but their intensity is weakened. To obtain high intensity data, lengthening the distance of the VSS and broadening the vertical divergence angle of the beam destroys the symmetry of the diffraction line profile: asymmetric diffraction line profile is obtained. This effect, called the **umbrella effect**, manifests as a skirt on the lower angle side of the diffraction line (diffraction angle 2θ below $90°$) and produces that at the higher angle side of the diffraction line (above $90°$). Take a careful look at the powder X-ray diffraction pattern of Figure 2.2, especially, considering the scan with a wide rectanglar slit. The center of the diffraction cone, of which diffraction angles are less than $90°$, is at $0°$ of the diffraction angle. While, the center of the diffraction cone, of which diffraction angles are above $90°$, is at $180°$. This example indicates the real situation in such a measurement system.

Figure 7.3 illustrates the umbrella effect in the zeolite LTA measured by the CuKα line. This material has a large lattice parameter, so the diffraction line appears at the lower-angle side and the umbrella effect is prominently displayed. As shown in the figure, the umbrella effect changes with the vertical divergence angle of the VSS. In typical optics, the divergence angle of the VSS is $2.5°$. At $0.5°$ divergence angle, the vertical divergence becomes small and the umbrella effect is suppressed, improving the symmetry of the profile. At present, the parameters describing optics be input to computer software that reproduces the resolution function. This technique, called the fundamental parameter method, has been encoded in the analysis program PDXL.

7.3.1 Slit width and resolution

The resolution of the equipment is decided by the optics of the diffractometer. In practical terms, the resolution is often determined by the

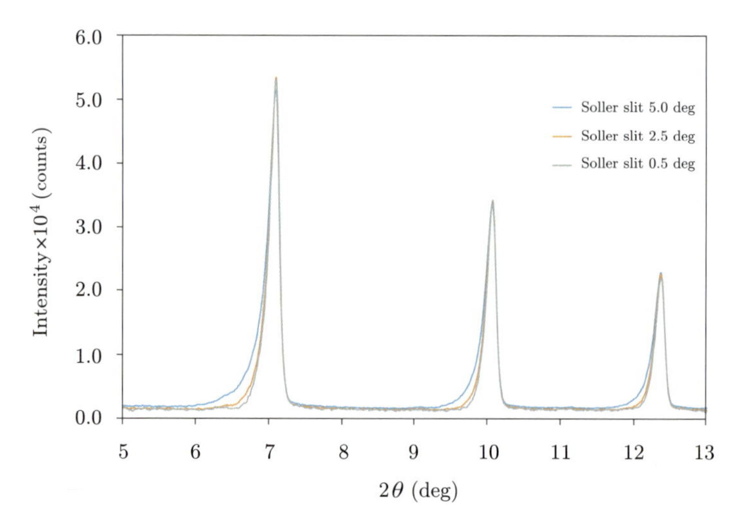

Figure 7.3 Diffraction pattern of zeolite-LTA based on the B-B method. Bragg reflections observed at low scattering angle are of the asymmetric intensity profiles having tail always in low scattering angle side but those at high scattering angle show a similar asymmetry having tail in high angle side. This is referred to as umbrella effect. The asymmetry depends on the divergence angle of the VSS: in turn with black, red, blue curve with decreasing the divergence angle.

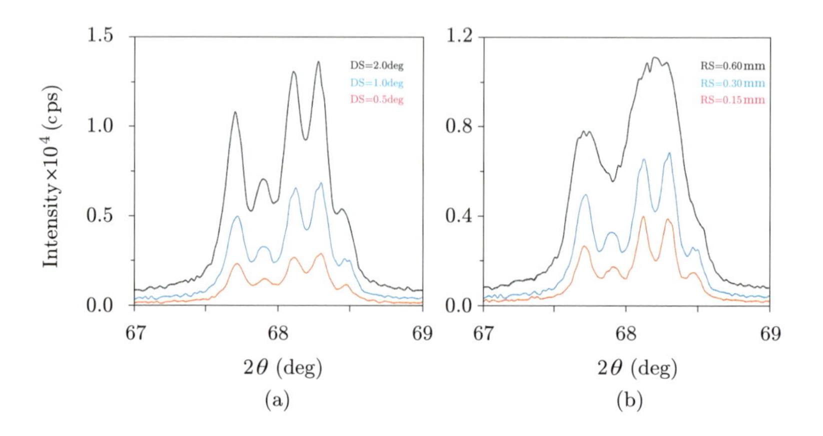

Figure 7.4 Relationship between the width of slit and the resolution.

separation between adjacent diffraction lines. As mentioned in Section 3.6, quartz irradiated by the CuKα line shows **five fingers** at diffraction angles between 67° and 69°. The diffraction lines overlap because they are closely spaced. The six diffraction lines arise from the separation of Kα_1 and Kα_2 at these diffraction angles, but only five peaks appear because the Kα_2 line of the (2 0 3) lattice plane overlaps the Kα_1 line of the (3 0 1) lattice plane. Here, we illustrate the changes in the five overlapped lines as the slit width is varied.

Figure 7.4 compares the resolutions at different slit widths. In the measurement, the DS and RS widths are changed by utilizing a goniometer with a camera length of $R = 185\,\mathrm{mm}$, instead of employing MiniFlex 300/600. We investigated the resolution of the optics by observing the presence of the five overlapped diffraction lines and their separation.

Figure 7.4 (a) shows that the profile changes as the divergence angle of the DS at the incidence side is widened to 0.5°, 1.0°, and 2.0°. Panel (b) shows that the profile changes for equivalent divergence angles of the RS at the receiving side. In both cases, widening the slit width degrades the resolution as expected but increases the intensity. The diffraction lines include the width from the sample. Therefore, rather than narrowing the DS and RS widths, we should focus on reducing the intensity. At a camera length of 150 – 180 mm, the RS is recomended to use 0.3 mm. In MiniFlex 300/600, the RS and R are fixed at 0.3 mm and 150 mm, respectively. When $R = 285\,\mathrm{mm}$, the RS is 0.45 mm in the standard case.

7.3.2 Deviation of the specimen surface

The B–B method uses convergent beam optics. If the sample surface shifts up and down, the diffraction angle of the observed diffraction line is also shifted. Therefore, we examine the relation between a shift on the sample surface and a shift of diffraction angle 2θ. Here, an up–down shift refers not to the sample surface but to a deviation on the θ rotation

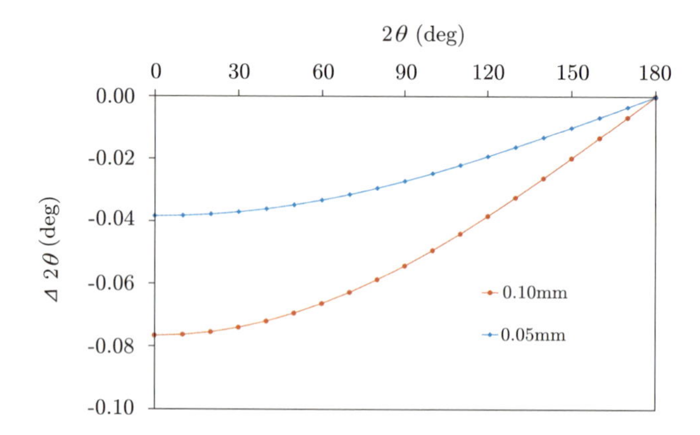

Figure 7.5 Shift of diffraction angle by deviation of sample surface. The shift $\Delta\ 2\theta$ is plotted against diffraction angle 2θ: the calculated results under the condition that the sample surface was deviated to upward direction by 0.05 mm and 0.10 mm. The camera distance is assumed to be 150 mm.

axis of the sample. A deviation of ΔL implies a diffraction angle shift of $\Delta 2\theta$. Given the camera length R, the shift is calculated as $\Delta 2\theta = -\tan^{-1}(2\Delta L \cos\theta/R)$.

Figure 7.5 shows the calculated $\Delta 2\theta$ for sample deviations of 0.10 mm and 0.05 mm. The magnitudes of $\Delta 2\theta$ are non-negligible especially at low diffraction angles. The difference between the $\Delta 2\theta$s reduces at high diffraction angles. In the precise measurement of the lattice parameters, we had better to select diffraction lines at high diffraction angle.

The shallow glass sample holder of MiniFlex 300/600 is 0.2 mm deep. If properly prepared, the sample surface will not deviate by 0.05 mm. However, even at 0.05 mm deviation, the angular error at the low-angle side will be at most 0.04°, which is within the allowance of sample identification (qualitative analysis). However, for accurate measurements of the lattice parameters, which are within MiniFlex 300/600's capabilities, the required angular precision is 0.01°. Therefore, careful packing of the powder sample is essential to avoid such deviation errors. To utilize the precision capabilities of MiniFlex 300/600, please pay special attention to the sample preparation.

7.4 Irradiated area of X-rays on the sample

7.4.1 Horizontal width

The width AB on a sample exposed by X-rays depends on the divergence angle of the X-rays (the DS slit width), camera length R, and incident angle θ. The width of the X-rays on the sample position under low-angle diffraction conditions ($2\theta < 20°$) is more than three times the width at high-incidence angles. When the sample is narrower than this width, the diffracted X-rays become weaker proportionally to the fraction of the incident X-rays on the sample. Therefore, more care is required for the X-ray diffraction with high angle. The incident X-rays which is not used must be calculated in advance. To this end, we calculate the width AB in MiniFlex 300/600, with a camera distance R of 150 mm and standard DS slits of 1.250° and 0.625°. The AB width is plotted as a function of diffraction angle 2θ in Figure 7.6.

The width of the standard sample plate is 20 mm. When DS = 1.250°, the sample is wider than the X-ray irradiation width on the sample surface if the diffraction angle 2θ exceeds 20°. If the diffraction angle 2θ is below 20°, the irradiation width is broadened and care is needed. On the other hand, when DS = 0.625°, the intensity of the diffraction pattern remains reasonable for over a 10° range of 2θ (Figure 7.6).

7.4.2 Vertical width

Like the horizontal width, the vertical width of the X-rays irradiated on the sample is an important factor. The irradiated vertical width is decided by the optics arrangement and is independent of the diffraction angle, so is relatively easy to handle. Figure 7.7 is a top view of the Soller slit VSS deployed at the incident side of the optics. The focusing length is defined by XX′, the focusing position of the X-ray source. Both the focusing length and the vertical width of the DS slit are usually fixed at 10 mm. Between X and DS, where the VSS is located, the vertical width of the beam is constant. After passing through the DS slit, the vertical beam broadens only within the range of the divergence angles permitted by the Soller slit. This phenomenon is detailed in Figure 7.7. The intensity is uniform within 10 mm of the beam center. In the figure, the spreaded vertical edge is the portion diverged through the Soller slit.

7.4.3 Penetration depth

Because every sample absorbs the X-rays, the penetration depth of the X-rays must be determined in the analysis. As explained in Section 1.5, the intensity I is given by $I_0 \exp\{-\mu x\}$, where x is the distance penetrated by

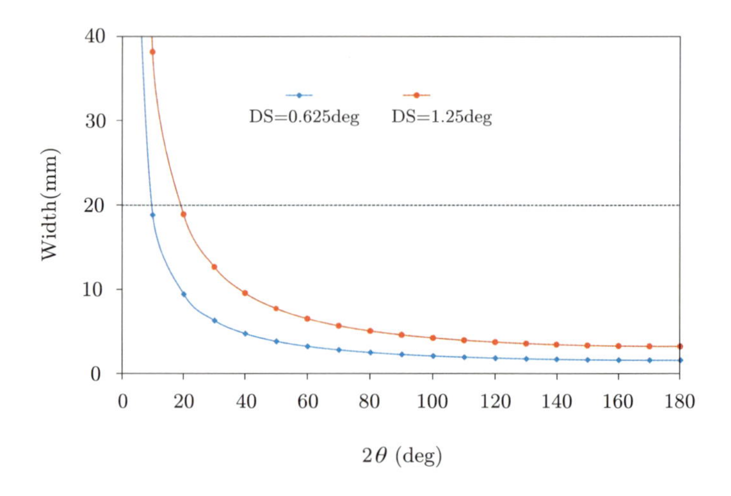

Figure 7.6 The width on a sample surface irradiated by X-rays against diffraction angle. The camera length is 150 mm.

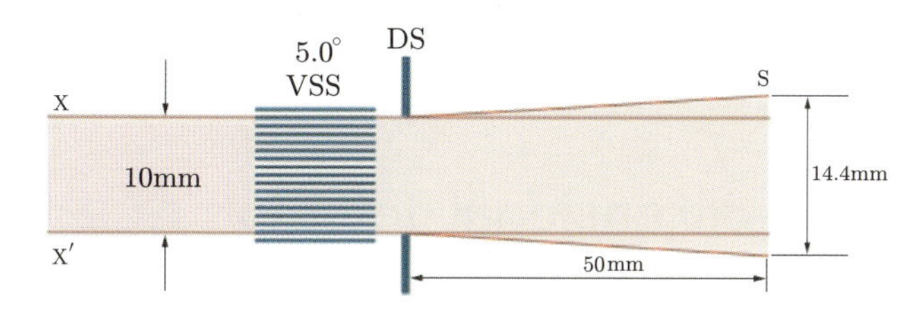

Figure 7.7 Vertical width of the X-rays

the X-rays through a material with linear absorption coefficient μ and I_0 is the non-attenuated intensity. Note that I_0 decreases exponentially with increasing x. Similarly, in the symmetrical diffraction case the intensity of X-rays reflected from a surface reduces by $\exp\{-2\mu t/\sin\theta\}$, where θ is the incidence angle on the sample surface and t is the penetration depth. Diffraction lines of low anlges are reflected from shallow, wide-area of the sample (near the surface), whereas those of high angles are reflected from deeper zones. Therefore, you should be carefully treated the data, when you analyze the diffraction pattern of the sample with density change in the thickness direction, or of the thin sample.

It goes without saying that samples with small absorption coefficients, which easily transmit X-rays, should be as thick as possible. The required thickness can be estimated from the diffraction patterns of a sample holder made from Al or some other well-known material. If the obtained data includes the diffraction line of the sample holder material, the sample is probably too thin. An non-reflective sample holder or a glass sample holder is recommended to avoid the background scatter in the qualitative analysis. It would be, however, better to avoid the reflections of these smaple holders for qualitative analysis.

In reflection diffractometry, the thickness is usually judged sufficient if $\mu t \fallingdotseq 3$. On the other hand, when observing diffraction patterns by the transmission method, the suitable thickness is about $\mu t = 1$. This estimation of thickness is based on that the intersity of transmitted X-rays becomes $1/e$.

7.5 One-dimensional detector

Recall that the B–B optical system uses the convergent beam method. In an ideal B–B system, a one- or two-dimensional detector requires no special preparation. However, these detcctors measures the profile over a certain width, not at one point. The profile, which is a function of 2θ, depends on the precision of the diffractometry as well as the aberration introduced by the measurement method. As mentioned earlier, this profile is called the resolution function. The diffraction line from a sample has a characteristic width, and thus, is certainly observed with the width. Even when the diffractometer is slightly deviated from the position of the Bragg reflection, the Bragg reflection will be captured by the resolution function.

If the RS slit in the diffractometer is replaced by a one-dimensional digital detector at the same position, the changing profile can be monitored. The signal that should be recorded as data is known at each position of the

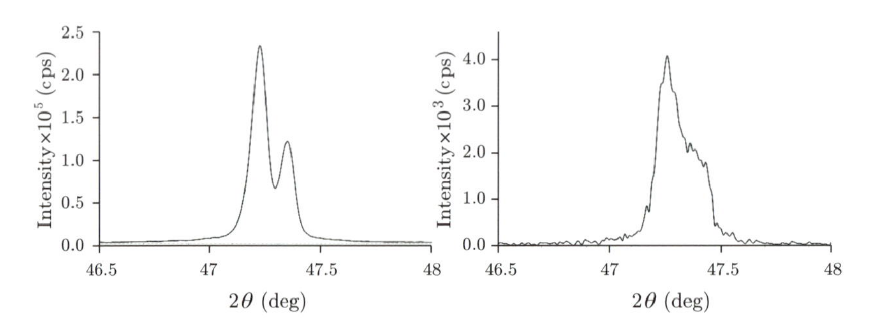

(a) By using one dimensional detector

(b) By using SC and RS slit

Figure 7.8 (2 2 0) reflection of Si.

detector. Such a test will provide a profile at each step; moreover, all of the recorded data are meaningful, even in convergent beam optics such as B–B, and none should be thrown away. In addition, when the intensity obtained at every step is recorded as a function of 2θ, the results correspond with profiles measured by detecfor through RS. The profiles obtained from one dimensional detector, D/teX Ultra and SC are compared in Panels (a) and (b) of Figure 7.8, respectively. Here, the peak shows the (2 2 0) reflection of a Si sample. Although both data were obtained by θ–2θ scanning, the profile is clearly better separated in Panel (a), showing better resolution than (b). This figure mathematically demonstrates the convolution [1]. The profile shows the convolution of resolution function for device with the diffraction line observed from the sample.

The D/teX Ultra detector is a highly sensitive semiconductor detector that captures the strong incident X-rays from minute areas. It operates by repeatedly and rapidly measuring the X-rays scattered at 2θ, and also by integrating the measurements [2].

The diffractometer installed with this detector should operate with the same precision as that of the detector. The precision of the 2θ arm of MiniFlex 300/600 is $0.02-0.04°$ per step. The D/teX Ultra detector is a one-dimensional detector with 128 channels, each with a pixel size of $0.1\,\mathrm{mm} \times 20.0\,\mathrm{mm}$. The position resolution is $\tan^{-1}(0.1\,\mathrm{mm}/150.0\,\mathrm{mm}) = 0.038°$ per pixel, equivalent to the device precision. On the other hand, in the Panel (b) of Figure 7.8, the RS slit width is 0.3 mm, corresponding to 0.11°. It means that the X-rays scattered from a sample are measured with the very narrow step of $0.02-0.04°$ by using the rather wide slit width of 0.11°.

7.6 Monochromatization of X-rays

The slit improves only the parallel property of the X-ray beam. The energy spectrum, which shows the wavelength distribution of the incident X-rays (Figure 5.3 (a)), is unaffected by the slit. The X-ray diffraction optics (except in special measurement methods) are designed to select X-rays of a certain wavelength. Conventional X-ray tubes using a metal target generate a relatively strong characteristic X-rays and a weak continuous X-rays. However, precise analysis of diffraction data is disturbed by poor P/B ratio with continuous X-rays and by unselected wavelength with other characteristic X-rays such as the Kβ line. Thus, when handling non-monochromatic X-ray beams, we must determine the wavelength dependence of the intensity by **spectral analysis** or by checking the spectrum performance. The extraction of one wavelength from X-ray spectrum is called **monochromatization**. The following sections described the X-ray monochromator and the filtering which are commonly used for the X-ray monochromatization.

7.6.1 Crystal monochromator

The most common spectral analysis method is based on the Bragg reflection of crystal, expressed as $\lambda = 2d \sin \theta$. On each lattice plane, the Bragg angle θ depends only on the wavelength λ, as d is fixed. Consequently, the spectrum of the incident X-rays is obtained by measuring the intensity change of the Bragg reflection using the diffractometer. Polycrystalline material is not appropriate as the monochromater, because it becames difficult to analyzed the spectram from the overlaped diffraction peaks. To avoid this problem, a single crystal is used with its lattice plane parallel to the surface of the sample holder. It sholud be careful because the higer-order Bragg reflections oppear, even in the single crystal, in the data.

A single crystal is set and its orientation is adjusted in such a way that CuKα X-rays are only obtained. This process is one of the processes of monochromatization, and the single crystal employed is called the **monochromator**. In general, the monochromator can be located at two positions: behind the X-ray source or before the the detector. In the first case, the incident X-rays are parallel and the Bragg reflection arises solely from the characteristic X-rays. In the second case, the X-rays diffracted from the sample are monochromatized before reaching the detector and this is called a counter monochromator method. Both methods remarkably reduce the X-ray intensity, and the reduced rate depends on the quality of single crystal. Monochromator crystals include highly oriented pyrolytic graphite (abbreviated as PG or HOPG), LiF, quartz, Si, and Ge. Recently, an **artificial multilayer** monochromator has been developed[21].

High-quality Si or Ge crystals produce high-quality parallel beam. By setting two or three single crystals on the path of incident X-rays, the parallelity of the beam is improved by two or three times of Bragg reflections. However, the intensity decrement cannot be avoided.

Figure 7.9 shows the arrangement of the powder X-ray diffractometer, which detects monochromatic X-rays by a monochromator (M) placed before the counter (SC). The graphite monochromator can be installed in front of the SC but not the D/teX Ultra due to the limited space. However, the D/teX Ultra operates with sufficient energy resolution (25 %) without the monochromator, as mentioned in Section 6.9. Generally, monochromatization reduces the intensity to about half that of the filter method (described next).

7.6.2 X-ray filters

The simplest monochromatization method is the Kβ filter method. As mentioned in Section 5.5, all elements exhibit an X-ray absorption edge. If the wavelength of the incident X-rays is shorter than the absorption edge, the absorption coefficient of the element alters by more than one order of magnitude. This principle underlies the Kβ **filter method**, in which the

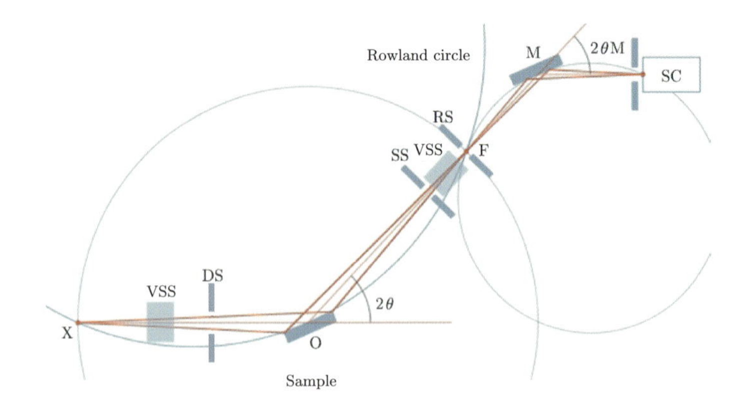

Figure 7.9 Optics with monochromator. The diffracted X-rays from the sample are converged on the focusing point F in this figure. After that, only Kα X-rays selected by monochromator crystal M is detected by SC.
M is called the counter monochromator.

flitering material absorbs X-rays of wavelength Kβ, close to the wavelength of the characteristic Kα X-rays. The filtering material is placed between the collimater and the sample or in front of the detector. The Kβ method absorbs the Kβ line and passes only the diffraction line of the Kα X-rays to the detector. Such methods are called filter methods or filter techniques. In order to exfract only the Cukα line from the Cu target, Ni filter is used, and Zr filter is used for the MoKα line.

As a concrete example, we will apply the Kβ filter method to Si powder. The absorption coefficient of Ni is larger for the CuKβ line (1.392 Å) than for the CuKα line (1.542 Å) because the absorption edge of Ni lies between the CuKα and CuKβ lines. When a thin Ni plate (thickness $\sim$0.015 mm) is inserted in the X-ray pass, the Kβ X-rays are attenuated to approximately 1/100 the intensity of the Kα line. Figure 7.10 presents the experimental evaluation of the Kβ filter. Panels (a) and (b) show the results of the Si powder sample in the absence and presence of the Ni filter, respectively. Comparing these results, we find that the Ni filter effectively blocks the

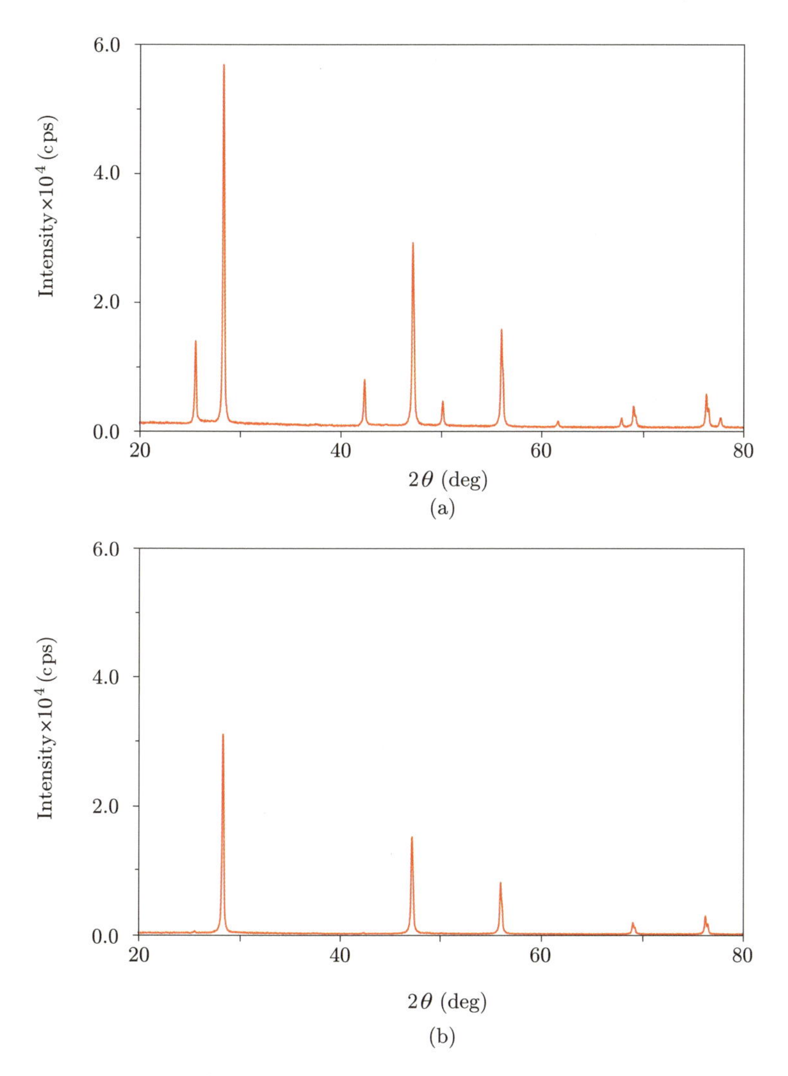

Figure 7.10 Role of the Kβ filter: The diffraction patterns from powdered Si. (a) is the pattern observed without Kβ filter; the diffraction lines by the Kβ line are observed. (b) is the data obtained with the Kβ filter, and the diffraction lines by the Kβ line are disappear.

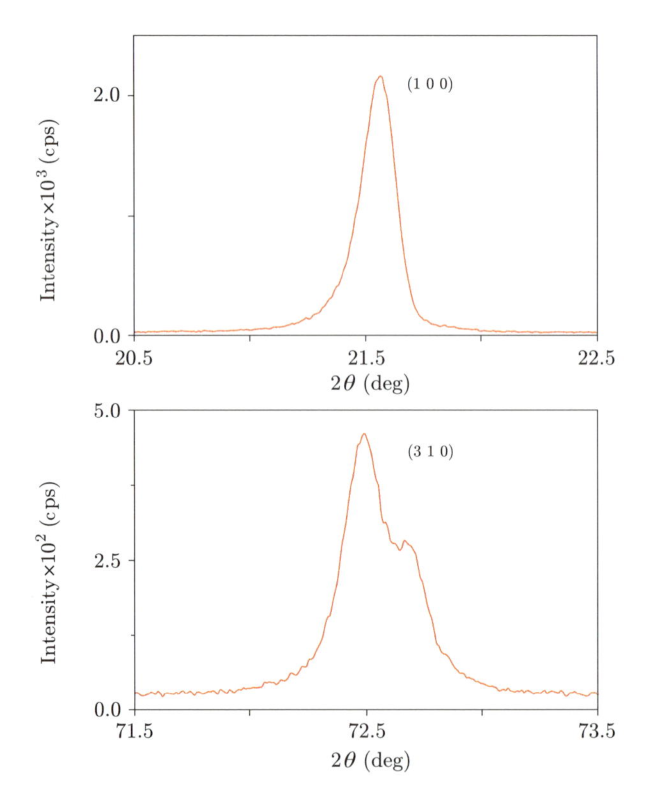

Figure 7.11 The change of diffraction profile with the increase of scattering angle, where the sample is CsCl. In the (1 0 0) diffraction line, the diffraction profiles by $K\alpha_1$ and $K\alpha_2$ line are overlapped. In the (3 1 0) diffraction line, two diffraction lines are starting to separate, near the diffraction angle of 72.5 degree; the intensity ratio of those two diffraction lines by $K\alpha_1$ and $K\alpha_2$ line is 2:1.

Cu$K\beta$ line.

7.6.3 Separation of $K\alpha_1$ and $K\alpha_2$

Figure 7.11 enlarges the two diffraction lines selected from the diffraction pattern of CsCl. The (1 0 0) Bragg reflection near 21.5° along the

2θ axis appears as one peak. However, the (3 1 0) diffraction line, with a higher index than (3 1 0), separates into two peaks indicating two independent diffraction lines. This result is expected because Kα X-rays with the strongest intensity generated from the X-ray target are actually a **doublet** (composed of Kα_1 and Kα_2). The intensity ratio $I\,(\mathrm{K}\alpha_1):I\,(\mathrm{K}\alpha_2)$ is $2:1$. The wavelengths of CuKα_1 and CuKα_2 for a copper target are 1.540593 Å and 1.544414 Å, respectively; therefore, $\Delta\lambda$ and $\Delta\lambda/\lambda\,(\mathrm{K}\alpha_2)$ are 0.003821 Å and 2.4×10^{-3}, respectively. The wavelength difference corresponds to a Bragg angle seperation of $\Delta\theta = \sin^{-1}\{(\Delta\lambda/\lambda)\tan\theta\}$. When θ is 70°, $\tan\theta$ is 2.75 and $\Delta 2\theta$ becomes 0.78°. The peak separation should always be considered for the diffraction line of higher order.

Annotations

1) The RS of the counter arm is replaced by a one-dimensional detector of a certain width, and scanning is performed by the θ–2θ method. At each step, the intensity data acquired at 2θ are stored in the memory server. This idea was suggested by H. Gobel at the end of the 1970s, but was not implemented because digital devices were very expensive at that time. One- and two-dimensional semiconductor pixel-type detectors were first deployed in the 2000s and have been reconsidered and subsequently used in diffractometry.

2) According to the RIGAKU catalog, the intensity of D/teX Ultra detector is sensitive more than 100 times that of SC detector. The D/teX Ultra retains all significant signals without the extra X-rays and reduces the measurement time.

Chapter 8 Error and data evaluation

The previous chapter explained the basic experimental tools of MiniFlex 300/600 as plainly as possible. This chapter focuses on the reproducibility and error of the obtained data. Understanding the precision of the experimental data is important for evaluating the analyzed result. This chapter reviews the theory of error estimation and considers a problem peculiar to the powder X-ray diffractometer, which may be already familiar to students of experimental diffractometry.

8.1 Error

When measuring a physical quantity, we should also estimate the **error** in the measured value. In this sense, what is an error? The error ε_j is defined as a difference between the true value X and the measured value X_j: $\varepsilon_j = X_j - X$. The subscript j indicates that ε_j is the j-th error among several measurements. However, as the true value is unknown (otherwise we would not need to measure it), ε_j cannot theoretically understand the error. Instead, we assume that the measured values vary around the true value. In many cases, the measured values follow a Gaussian distribution around the central value (it is also called the normal distribution), as shown in Figure 8.1. Therefore, we calculate the arithmetic mean $\langle X \rangle$, or **average**, of the measured values. Then, instead of ε_j, we compute the deviation of each measured value from the average $\Delta_j = X_j - \langle X \rangle$ and the **standard (root mean square) deviation**:

$$\sigma = \sqrt{\frac{\sum \Delta x_j{}^2}{n(n-1)}} \tag{8.1}$$

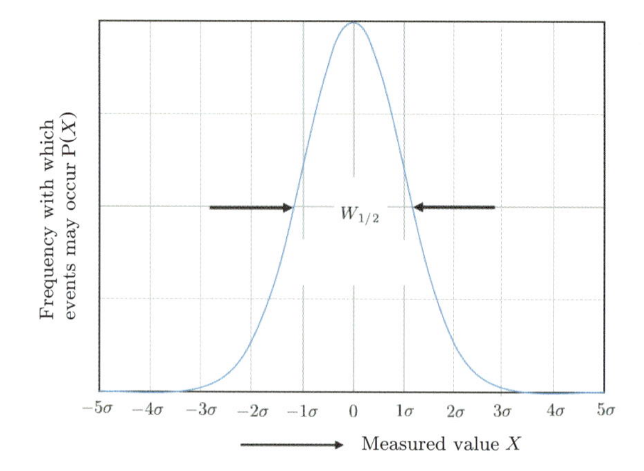

Figure 8.1 Gaussian distribution. The measured value shows the Gaussian distribution around the value that is the most reliable. $W_{1/2}$ is the FWHM, and is 2.35 times of the standard deviation σ .

The **percentage of standard deviation**, also called the relative standard deviation, is then calculated as follows.

$$\sigma(\%) = \left(\frac{\sigma}{\langle X \rangle}\right) \times 100 \tag{8.2}$$

Above argument is the error theory due to the Gaussian distribution. In this theory, the obtained physical quantity is generally written as $X = \langle X \rangle \pm \sigma$. An experimental physical quantity, or a theoretical calculation based on experimentally measured quantities, is expressed in this notation. $W_{1/2}$ in Figure 8.1 denotes the width of the Gaussian distribution, equal to 2.35σ.

Here, we have discussed accidental errors in above; errors not caused by artificial phenomena. Such errors are possible to deal with statistical treatment. In addition, there are systematic errors which are introduced by insufficient handling of equiments. Both **accidental error** and **systematic errors** are important in diffraction instruments. A concrete example with respect to systematic errors is presented in the later section.

8.2 Reproducibility

8.2.1 Peak position

For identifying the structure of a material, we prefer to measure as many diffraction lines as possible. However, the diffraction lines measured within a range of diffraction angle given by MiniFlex 300/600 are sufficient. As mentioned in Section 1.4, an identification is confirmed when the lattice parameter from the sample is calculated from each diffraction line with no contradictions. The result might not necessarily agree with the reported value of the lattice parameter, but the size of the unit cell must be determined with sufficient accuracy. Therefore, the lattice parameter is calculated with some uncertainty from numerous diffraction lines, and not by measuring the accurate diffraction angle of one diffraction line. However, when strain measurements of a given sample are required, we must precisely measure the diffraction angles of special reflections.

We are often asked whether MiniFlex 300/600 has sufficient angular precision to accomplish such measurements. Actually, MiniFlex 300/600 has the same precision as that of other diffractometers. As written in the instruction manual, we measure as many diffraction lines as possible by using a standard reference material whose lattice parameters are known with high accuracy, such as powdered Si; based on these data, the diffraction angles measured can be corrected. In other words, the precision of the instrument is of very high because the measured angles are automatically corrected in reference to calibration curve stored in the computer connected to the MiniFlex 300/600.

The angle measurements require attention. The diffractometer measures the diffraction line at various incidence angles θ and scattering angles 2θ of the X-rays, which are regularly spaced by a specified step size. However, not only this machine but all the diffractometer use gear drive, which inevitably introduces mechanical backlash when the gears engage. Without such backlash, the machine cannot work exactly. When the gear reverses direction immediately after a revolving operation, an immovable area, called play, is created, causing error in the angle measurement. For instance, suppose that the diffraction lines are measured from the low angle side to the high angle side, and then suddenly measured from the high angle side to the low angle side for confirmation. In this case, the play area would cause a difference between the measurement sets. Therefore, when checking data by repeated measurements, it is a regulation to maintain the same direction.

8.2.2 Intensity

When X-rays are detected, measured value is always with in statistical fluctuations. If the value obtained by detector is N, the magnitude of fluctuation is given by $\sqrt{N}$. The measured value of X-rays is therefore $N \pm \sqrt{N}$. $\sqrt{N}$ is merely the standard deviation σ derived by the error theory reviewed in the previous section.

When the standard deviation calculation (Equation (8.1)) is repeated several times under slightly changing conditions, the $\sqrt{N}$ values will not coincide, demonstrating that errors other than statistical one result from systematical reasons. Here, let us distinguish between the standard deviation of a measurement $\sigma_{\text{ideal}}(= \sqrt{N})$ and that calculated by Equation (8.1), σ_{obs}. The standard deviation is contributed by σ_{ideal} (%) and σ_{obs} (%), where σ_{ideal} (%) is expressed as $\sqrt{N}/N \times 100 = 1/\sqrt{N} \times 100$. Both these standard deviations are well defined, but readers should carefully consider their appropriateness in a particular application.

In any case, the number of measurements should be increased to lower the standard deviation. For example, to measure the integrated intensity of a certain diffraction line within 1% error, 10,000 integrated intensity counts are required. When the scanning speed (which controls intensity) of the diffraction instrument is changed, the total count changes accordingly; in order to examine the "fluctuation", we measure the integrated intensity of the (0 1 2) diffraction line using an Al_2O_3[1)] (hexagonal) powder sample. Measurements are performed at scanning speeds of 20°/min and 4°/min (10 measurements per each). The results are summarized in Table 8.1.

The first row shows the number of measurements j. The second and third rows show the integrated intensities N_j measured at 20°/min and 4°/min, respectively. The data in the fourth row were measured 10 times at 4°/min with Al_2O_3 powder repacked in the sample holder at each measurement. The last row states the average value $\langle N \rangle$, $\sigma_{\rm obs}$, and $\sigma_{\rm obs}$ (%) of each line. The error range of the measurements is indicated on the right of each measured value; no mark indicates that the data is within $\langle N \rangle \pm \sigma_{\rm ideal}$, whereas "$\langle 2\sigma_{\rm ideal}$" or "$\rangle - 2\sigma_{\rm ideal}$" indicates that the data is below or above

Table 8.1 Reproducibility of the integral intensity and the repacking of Al_2O_3 powder.

	Reproducibility of the measurement		Reproducibility by the repacking
	(20°/min)	(4°/min)	(4°/min)
1	5468 $> -2\sigma_{\rm ideal}$	27435	27260 $> -2\sigma_{\rm ideal}$
2	5642	27165	28009 $< 2\sigma_{\rm ideal}$
3	5587	27091	28051 $< 2\sigma_{\rm ideal}$
4	5479 $> -2\sigma_{\rm ideal}$	27289	27977
5	5464	27552 $< 2\sigma_{\rm ideal}$	28002 $< 2\sigma_{\rm ideal}$
6	5508	27460	27517 $> -2\sigma_{\rm ideal}$
7	5625	27339	28058 $< 2\sigma_{\rm ideal}$
8	5710 $< 2\sigma_{\rm ideal}$	27381	27469
9	5792 $< 2\sigma_{\rm ideal}$	27498 $< 2\sigma_{\rm ideal}$	27868
10	5635	26944 $> -2\sigma_{\rm ideal}$	28094 $< 2\sigma_{\rm ideal}$
$\langle N \rangle$	5591	27315	27831
$\sqrt{\langle N \rangle}$	74	165	167
$\sigma_{\rm obs}$	111	185	284
$\sigma_{\rm obs}$(%)	2.0	0.7	1.0

the value of $\langle N \rangle \pm 2\sigma_{\text{ideal}}$.

This result provides us with considerable information. In the second line (20°/min), $\sigma_{\text{ideal}} = \sqrt{N}$ is 74, whereas σ_{obs} is 111; σ_{obs} is larger than σ_{ideal}. We see that 6 out of 10 measurements are within the range $\langle N \rangle \pm \sigma_{\text{ideal}}$. However, even such 4 measurements do not cross over the range $\langle N \rangle \pm 2\sigma_{\text{ideal}}$. Although σ_{obs} is large, it closely approximates an ideal Gaussian distribution (Figure 8.1) because the measuement reproducibility is mostly within the range σ_{ideal} and the difference is symmetrically distributed over $\pm 2\sigma_{\text{ideal}}$.

In the third line, σ_{ideal} (165) is close to σ_{obs} (185), comparing with that of second line. In addition, 7 out of 10 measurements were in the range $\pm\sigma_{\text{ideal}}$, approaching the ideal distribution. As the measurement value shows 5 times intensity of second line, the standard deviation falls by 1/3 from 2.0 % to 0.7 %. Thus, the precision improves.

Line 4 checks the reproducibility of repacking the sample. Although the average value is similar to that in Line 3, the standard deviation σ_{obs} is 284, far from the ideal σ_{ideal} of 167. In addition, only 3 out of 10 measurements were within $\pm\sigma_{\text{ideal}}$; 7 measurements were distributed beyond $\pm\sigma_{\text{ideal}}$ and were inclined to the $+2\sigma_{\text{ideal}}$ side. This indicates the presence of systematic error in addition to simple statistical fluctuation in the measurements. Repacking the sample diverges the standard deviation from its ideal value. Therefore, the repacking work needs careful attention.

8.2.3 Background

When observing and analyzing a diffraction pattern, background processing is needed for convenience of data processing. Figure 8.2 shows a general case of diffraction line, where the measured background differs between low angle side and high angle side of the diffraction peak[2)].

In Figure 8.2, the measured values are N_1 and N_2 at angles $2\theta_1$ and $2\theta_2$, respectively, where $2\theta_2$ is positioned after m steps from $2\theta_1$. A diffraction

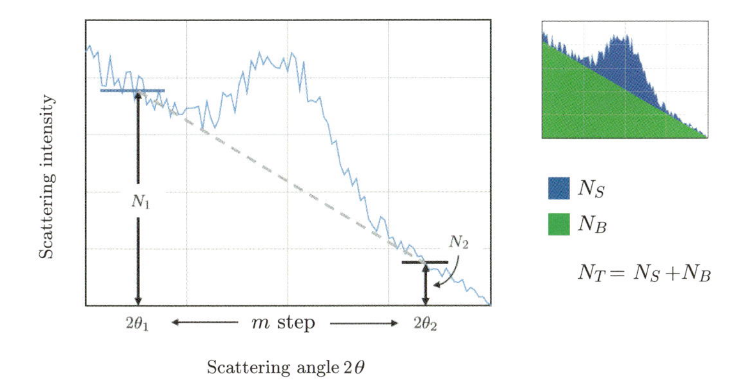

Figure 8.2 Diffraction line with high backgroud.

peak appears between $2\theta_1$ and $2\theta_2$. The total measured background N_B between N_1 and N_2 is assumed as $N_B = \{(N_1 + N_2)/2\} \cdot m$. On the other hand, the integrated value N_S of the diffraction line includes the measurements acquired from $2\theta_1$ to $2\theta_2$ along with the scattering angle. After summing up the measured values as $N_T\{= \Sigma N_j(j = 1......m)\}$, N_S is obtained by subtracting N_B from N_T, yielding Equation (8.3).

$$N_S = (N_T - N_B) \pm \sqrt{N_T + N_B} \tag{8.3}$$

When N_1 and N_2 are equal and diffraction line is symmetric with respect to its center, the background is easily computed as $N_B = mN_1$. In addition, when N_1 is quite low, the background N_B in the total volume is smaller than the standard deviation calcualted from the integrated signal value; that is, $N_B < \sqrt{N_S}$. In this case, the background can be ignored.

8.3 Condition of powder specimen

In the example of Table 8.1, repacking the Al_2O_3 powder alteres the intensity of the diffraction lines. The resulting standard deviation cannot be completely explained by background radiation. This phenomenon is a well-known results obtained by using the sample with non-uniform crystalline sizes. It is easily understood if the diffraction patterns of such a sample are observed by two-dimensional detector. The imaging plate equipped in the RINT-Rapid diffractometer (Section 2.1) is suitable for this purpose.

Figure 8.3 shows part of diffraction rings observed by RINT-Rapid. The average particle sizes are 3 μm, 7 μm, and 10 μm. The powder sample was filled flatly into the sample holder. We focus on a region of diffraction ring near the equator line (dash-dotted line). Large particles produce a speckled diffraction ring. At smaller particle sizes, the small flecks connect to form

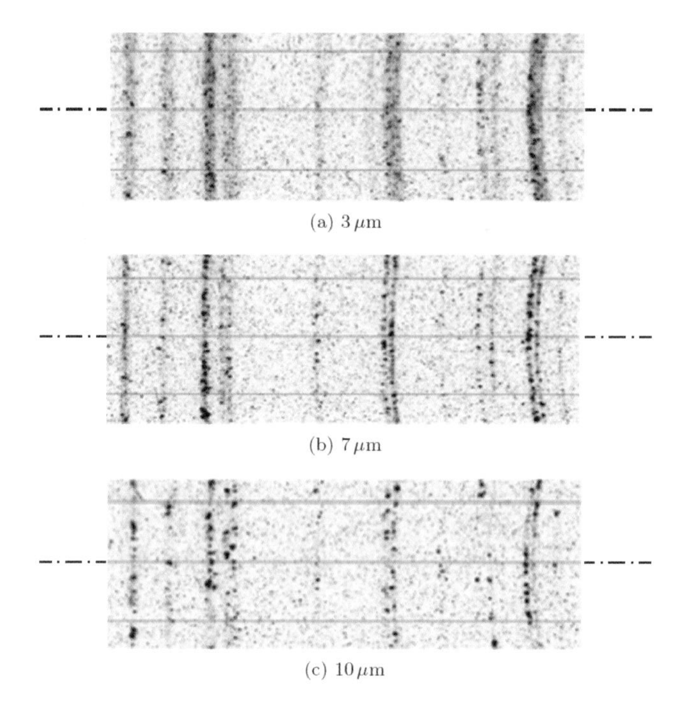

(a) 3 μm

(b) 7 μm

(c) 10 μm

Figure 8.3 Change of powder X-ray diffraction ring by different particle size of Al_2O_3. It is measured by RINT-Rapid.

rings.

Diffraction pattern observed by diffractometers such as MiniFlex 300/600 are corresponded to part of diffraction rings near the equator line. Repacking a sample changes the sizes and numbers of particles in the specific area of the powder X-ray ring. The standard deviation σ_{obs} differs from $\sqrt{\langle N \rangle}$, as explained in Section 8.2.2. As the particle size reduces, more particles occupy the irradiated area and contribute to the diffraction.

Therefore, to improve the intensity reproducibility of the diffraction line, the particle size of the sample should be reduced in a mortar. If no existence of single grain was confirmed by touching between the fingers, the resulting powder should be sufficiently fine to eliminate any sense of graininess. Altough only briefly mentioned here, organic crystals must be crushed very carefully as they are easily broken by the crushing.

8.4 Rotation sample stage

A rotation sample stage rotating in the in-plane direction will increase the time average of the number of particles contributing to the diffraction. This method achieves almost the same result as crushing the sample. Panels (a) and (b) of Figure 8.4 show the diffraction patterns of a stationary

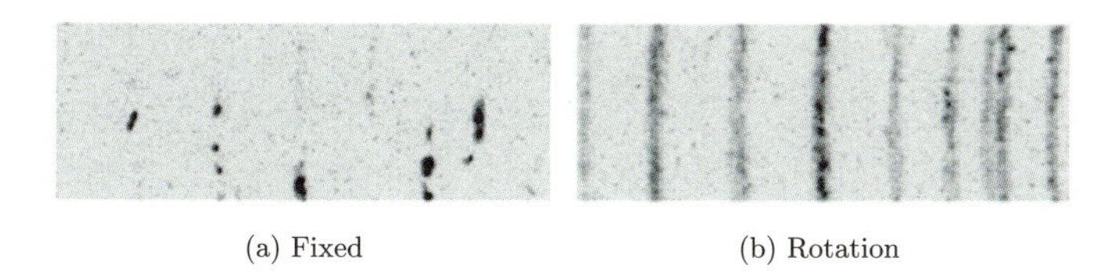

(a) Fixed (b) Rotation

Figure 8.4 Effect of inplane rotation of sample.
(a) is the diffraction pattern when the sample is fixed; showing a lot of spots so as not to be able to confirm a diffraction line.
(b) is observed with rotation of the sample stage; it shows a continuous diffraction line.

and rotated large-sized quartz sample, respectively. While only diffraction spots appear in the stationary case, rotating the sample stage in the in-plane direction yields connected diffraction rings similar to that of small particles. This demonstrates the effectiveness of the rotation sample stage.

The integrated intensities of a rotational and an irrotational method were compared for particles of various sizes. The reproducibility of the integrated intensities obtained by the irrotational method were insufficiently for the sample with the average particle size in $25-30\,\mu$m. On the other hand, the rotational method yielded good reproducibility for all particle sizes, and was thus suitable for quantitative analysis. In general, the sample should be ground to approximately $10\,\mu$m for quantitative analysis. This result serves as a useful reference.

Annotations

1) This is also called corundum. The crystal structure is hexagonal. Large single corundum crystals are rubies when red and sapphires when blue. The powder is used as abrasive.

2) In what scenario can we find the high background shown in Figure 8.2 (b)? Obviously, such a situation should be avoided if possible. Let us consider the following case. When the energy of the incident X-rays is higher than the energy required to excite the fluorescence X-rays of the element included in the sample, most of the X-rays are absorbed by the sample. Fewer of the X-rays contribute to the diffraction, and the intensity of the diffraction line weakens while the fluorescence X-ray background increases. Such a phenomenon occurs when we use the X-ray source of a wavelength causing a large absorption of main element in the measurement sample. For instance, the characteristic X-rays of CuKα and CuKβ are not suitable for investigating the structures of materials including Co, Fe, and Mn, because of the large μ/ρ for these elements.

Appendix

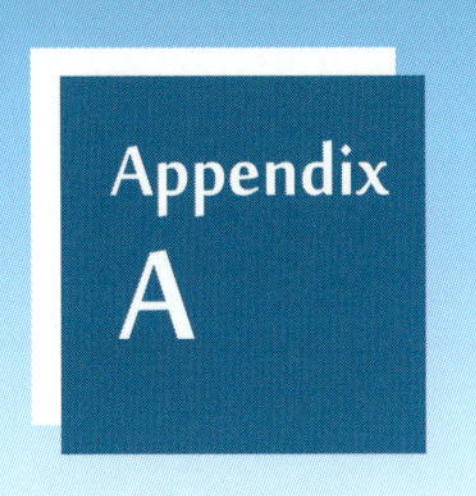

Basic concept of X-ray scattering and diffraction

In this Appendix, we present the basic concepts of X-ray scattering/diffraction by a crystal. As the reader already knows, crystals consist of atoms, ions, or molecules that repeat their relative locations in three dimensions. Therefore, all crystals are composed of repeating units called unit cells, in which atoms are arranged at specific positions. Thus, we need to understand how unit cells scatter X-rays in specific directions. First, we need to familiarize ourselves with X-ray scattering phenomena by atoms. However, since an atom is a collection of electrons, we also need a mathematical description of X-ray scattering from electrons. Therefore, our discussion begins with the scattering of X-rays from electrons, then progresses to scattering by atoms and crystals. Metaphorically, this route can be considered as the royal road to understanding the scattering and diffraction phenomena of X-rays by substances.

A.1 Plane and spherical wave

Plane wave: X-rays are a type of electromagnetic wave. Thus, we first require a mathematical description of electromagnetic waves. An electromagnetic wave is a transverse wave (defined as a traveling wave that oscillates perpendicularly to its direction of travel) consisting of two perpendicular components: an electric field $\boldsymbol{E}$ and a magnetic field $\boldsymbol{H}$, both of which are perpendicular to the propagation direction $\boldsymbol{s}$. $\boldsymbol{E}$ is expressed as a sinusoidal function, as described below for $\boldsymbol{E}(\boldsymbol{r}, t)$.

$$\boldsymbol{E}(\boldsymbol{r}, t) = \boldsymbol{E}_0 \sin\left\{-2\pi\left(\frac{\boldsymbol{sr}}{\lambda}\right) + 2\pi\left(\frac{c}{\lambda}\right)t\right\} \tag{A.1}$$

Where, $\boldsymbol{r}$, c, t, and λ denote the position vector, velocity of light, time, and wavelength, respectively. $\boldsymbol{E}_0$ is the peak magnitude of the wave. The electric field $\boldsymbol{E}$ represents the amplitude of the X-rays, and its square represents the intensity, which is proportional to the energy of the wave. As explained in the text, the intensity or square of the amplitude defines the probability of finding the X-ray photons. In this equation, $2\pi\boldsymbol{s}/\lambda$ indicates that the electromagnetic wave propagates in the $\boldsymbol{s}$ direction with wavenumber $2\pi/\lambda$. Thus, we denote $2\pi\boldsymbol{s}/\lambda$ as $\boldsymbol{k}$. The frequency ν is given by c/λ, which is the propagating distance per unit time divided by the wavelength. More simply, we can express the above Equation (A.1) in terms of the angular frequency ω instead of $2\pi c/\lambda = 2\pi\nu$.

$$\boldsymbol{E}_p(\boldsymbol{r}, t) = \boldsymbol{E}_0 \sin\{-\boldsymbol{kr} + \omega t\} \tag{A.2}$$

Equation (A.2) describes a plane wave moving in the direction of $\boldsymbol{k}$. From this equation, we can determine the amplitude of the electric field $\boldsymbol{E}$ at position $\boldsymbol{r}$ and time t. When dealing with scattering phenomena, we often encounter several overlapping waves with different phases. In such a calculation, it is mathematically convenient to convert the sinusoidal form of the plane wave (A.2) to the following complex exponential form.

$$\boldsymbol{E}_p(\boldsymbol{r}, t) = \boldsymbol{E}_0 \exp\{-i(\boldsymbol{kr} - \omega t)\} \tag{A.2$'$}$$

Which is related to (A.2) through Euler's identity, $\exp\{\pm ix\} = \cos x \pm i \sin x$.

Spherical waves: A spherical wave is a wave whose front spreads uniformly in all directions from a point source. The intensity $\boldsymbol{E}^2$ (or energy) of a spherical wave decreases as the inverse square of the distance from the point source. Thus, a spherical wave is described by

$$\boldsymbol{E}_s(\boldsymbol{r}, t) = \frac{\boldsymbol{E}_0}{r} \exp\{-i(\boldsymbol{kr} - \omega t)\} \tag{A.3}$$

where $\boldsymbol{k}$ and $\boldsymbol{r}$ are the wave vector and position vector, respectively, based on the coordinate of which origin is taken at location of the point source.

When X-rays (which are plane waves) collide with a stationary electron, Thomson scattering occurs. The scattered X-rays are expressed as a spherical wave (A.3) originating from the position of the electron.

A.2 Thomson scattering and polarization factor

Thomson scattering: Material is formed from aggregates of atoms, which themselves consist of many electrons and nuclei. Both electrons and nuclei are particles with masses and electrical charges, whereas X-rays are electromagnetic waves. Therefore, X-ray scattering by a material is considered to be the scattering of an electromagnetic wave by an aggregate of many charged particles.

Let us consider X-ray scattering by a single charged particle with mass M and charge e that is located at the origin. A plane wave of X-rays polarized in a specific direction in the x–z plane propagate in the positive direction along the y-axis. The electric field $\boldsymbol{E}$ of the X-rays displaces the charged particle from its origin, forcing it to oscillate. Therefore, the displacement of the particle by an oscillating electric field can be expressed by the equation of motion of forced oscillation. As the oscillation of a charged particle can be regarded as a dipole oscillation, we can easily imagine the emission of electromagnetic wave from that position. This simple scenario describes X-rays scattered from a charged particle.

As a result, the dipole oscillates in the same direction as the electric field $\boldsymbol{E}$ of the incident X-rays. Its magnitude is given by $\pm(e^2/Mc^2)$, and its frequency is that of $\boldsymbol{E}$, where $\pm$ represents the sign of the electric charge of the particle. After scattering by an electron (a negatively charged particle), the scattered wave is 180° (π radians) out of phase with the incident wave. This phase difference implies an inverse relation between the oscillating electron and the electric field vector. In the case of scattering by a nucleus (a positively charged particle), we must consider that the proton is 1823 more massive than the electron, that is, M =1823 m, where m is the mass of an electron. As a consequence of their much greater weights, positive nuclei contribute negligibly to the scattering, so the scattering can be regarded as wholly contributed by electrons.

The scattering amplitude of an electron (e^2/mc^2), which shows the classical electron radius, is of the order of 10^{-13}. This scattering phenomenon is called Thomson scattering after its discoverer, J. J. Thomson. Because the incident and scattered waves have the same frequency, Thomson scattering is a type of elastic scattering[1)].

Polarization effects: In the above argument, the incident X-rays were assumed to be polarized in a specific direction on the x–z plane in Figure A.1. Accordingly, the scattered waves are polarized in the same direction as the incident X-rays, but their intensity depends on the angle from which the scattering is observed. In other words, the intensity depends on the observed scattering angle. This effect is called the polarization effect of Thomson scattering.

Panels (a) and (b) of Figure A.1 show two extreme effects for different polarizations of the incident X-rays on the observed Thomson scattering. The scatterer (an electron) is located at the origin. In Figure A.1 (a), the incident X-rays are polarized along the z-axis; in panel (b), they are polarized along the x-axis. When the scattered waves from the origin are observed at a scattering angle of 2θ on the x–y plane, the amplitude of the Thomson scattering is e^2/mc^2. As shown in panel (a), this amplitude is independent of the observation angle 2θ. However, in panel (b), the amplitude of the Thomson scattering changes with 2θ as $(e^2/mc^2) \cdot \cos 2\theta$. Therefore, when viewed from $2\theta = \pi/2$, the scattering amplitude is zero, as if no scattering had occurred.

In general, the electric field vector $\boldsymbol{E}$, which travels in an arbitrary direction in the x–z plane, is given by two orthogonal components $\boldsymbol{E} = \boldsymbol{E}_{\perp} + \boldsymbol{E}_{=}$,

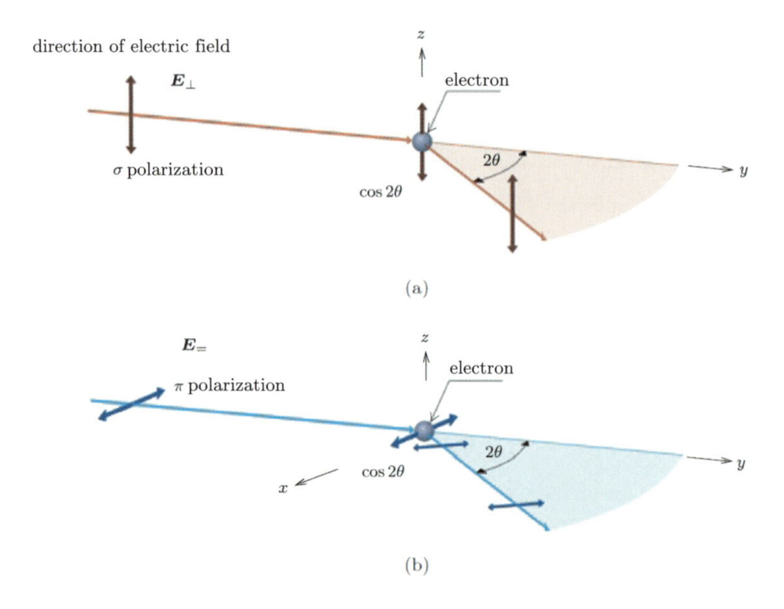

Figure A.1 Two different polarizations of electromagnetic wave.

where $\boldsymbol{E}_\perp$ and $\boldsymbol{E}_=$ are the perpendicular and parallel vector components, respectively, to the x–y plane, and $|\boldsymbol{E}_\perp| \neq |\boldsymbol{E}_=|$. These two components are related by $|\boldsymbol{E}|^2 = |\boldsymbol{E}_\perp|^2 + |\boldsymbol{E}_=|^2$, as is easily confirmed by Pythagoras' Theorem. This rule is obeyed by all electromagnetic waves. Thus, if the incident X-ray waves are polarized in some specific direction in the x–z plane, the intensity of the Thomson scattering waves is obtained by summing the intensities of the two orthogonal components.

$$I_s = \left(\frac{e^2}{mc^2}\right)^2 \{|\boldsymbol{E}_\perp|^2 + \cos^2 2\theta |\boldsymbol{E}_=|^2\} \tag{A.4}$$

As mentioned above, this expression is applicable only to incident X-rays polarized in a specific direction. However, the polarization direction is difficult to determine in general. Moreover, measuring the scattering X-rays with sufficient accuracy requires time. Consequently, all X-ray measurements are time-averaged measurements. As the incident

X-rays are considered to uniformly change their polarization from parallel to perpendicular during the measurement, we can take the averages of $|\boldsymbol{E}_\perp|^2$ and $|\boldsymbol{E}_=|^2$ as $\langle|\boldsymbol{E}_\perp|^2\rangle$ and $\langle|\boldsymbol{E}_=|^2\rangle$, respectively, which are related by $\langle|\boldsymbol{E}_\perp|^2\rangle = \langle|\boldsymbol{E}_=|^2\rangle = (1/2)\langle|\boldsymbol{E}|^2\rangle$. Thus, we have

$$I_s = \left(\frac{1+\cos^2 2\theta}{2}\right)\left(\frac{e^2}{mc^2}\right)^2 I_0 \tag{A.5}$$

In this expression, we have replaced $\langle|\boldsymbol{E}|^2\rangle$ with the intensity of the incident X-rays I_0. The factor $\{(1+\cos^2 2\theta)/2\}$ is called the **polarization factor** of the Thomson scattering power $(e^2/mc^2)^2$. If the distance between the scatterer and observation point is R, the intensity that can be detected through a solid angle $d\Omega$, which is inversely proportional to the square of R, is given by

$$I_s = \left(\frac{1+\cos^2 2\theta}{2}\right)\left(\frac{e^2}{mc^2}\right)^2 I_0\frac{d\Omega}{R^2} \tag{A.6}$$

Usually, the factor $d\Omega/R^2$ is not discussed, but is treated as a simple coefficient.

Above, we mentioned that e^2/mc^2 is the classical electron radius. Thus, $(e^2/mc^2)^2$ has the same dimensions as area. Its value is calculated as $(e^2/mc^2)^2 = (4.8\times10^{-10})^4/\{(9.1\times10^{-23})^2(3\times10^{-10})^4\} = 7.9\times10^{-26}$ cm^2, where we have used CGS units instead of MKS units. On the other hand, omitting the polarization factor from Equation (A.6), we find that $(e^2/mc^2)^2$ is the intensity ratio, I_s/I_0, of the scattered X-rays to the incident X-rays. In other words, the square of the classical electron radius is the scattering cross-section, describing the probability that X-rays will be scattered by the electron. At first glance, the cross-section (which is on the order of 10^{-13}) appears infeasibly small. However, in real X-ray scattering measurements, the vast number of electrons greatly increases the probability of X-ray scattering.

A.3 Phase difference and scattering vector

Figure A.2 shows two scatterers at locations P and Q, separated by a distance $\boldsymbol{r}$. If both scatterers are irradiated by a plane wave of wavelength λ propagating in the direction specified by unit vector $\boldsymbol{s}_i$, they emit a spherical scattering wave each. Let us further assume that we are observing the scattering waves at a point far from the scatterers, in the direction given by the unit vector $\boldsymbol{s}_f$. The path difference between the two waves is $\mathrm{P'Q} - \mathrm{PQ'}$, where the waves scattered at P and Q propagate in the same direction, $\boldsymbol{k}_f$.

As indicated in the figure, the wavenumber vector of the incident wave is

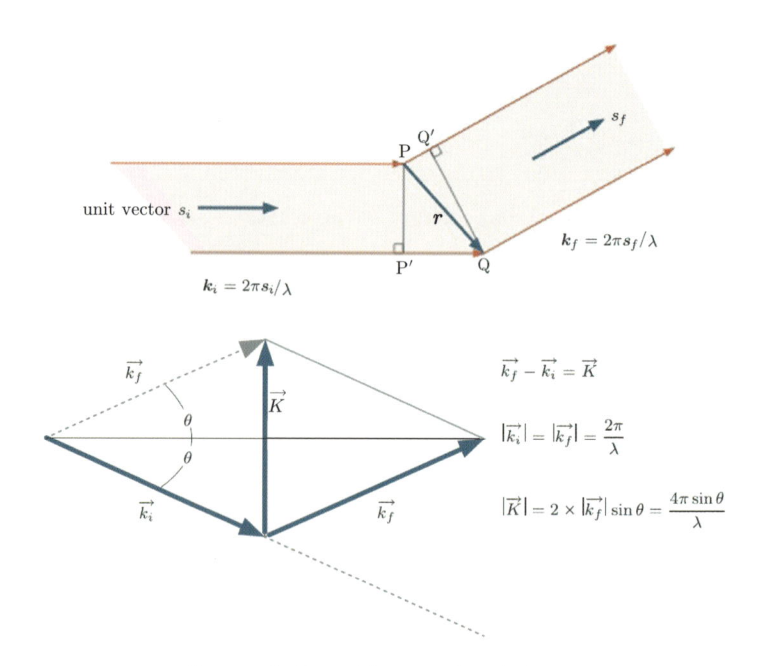

Figure A.2 Phase difference and scattering vector.

given by $\boldsymbol{k}_i(=2\pi\boldsymbol{s}_i/\lambda)$. When viewed from a sufficiently far distance, the spherical waves scattered from P and Q can be regarded as plane waves with wavenumber vector $\boldsymbol{k}_f$ $(=2\pi\boldsymbol{s}_f/\lambda)$. Let us now consider the path difference $\mathrm{P'Q}-\mathrm{PQ'}$ between the scattered waves from P and Q, both with the same $\boldsymbol{k}_f$. The phase difference ϕ is obtained by dividing the measured path difference by the wavelength λ, and multiplying this fraction by 2π. As shown in the figure, P′Q is the $\boldsymbol{s}_i$-directional component of PQ $(=\boldsymbol{r})$, so is given by the scalar product of $\boldsymbol{r}$ and $\boldsymbol{s}_i$. Thus, the path difference is given by

$$\begin{aligned}\mathrm{P'Q}-\mathrm{PQ'} &= (\boldsymbol{r}\boldsymbol{s}_i)-(\boldsymbol{r}\boldsymbol{s}_f)\\ &= \boldsymbol{r}(\boldsymbol{s}_i-\boldsymbol{s}_f)\end{aligned} \tag{A.7}$$

The quantities $2\pi\boldsymbol{s}_i/\lambda$ and $2\pi\boldsymbol{s}_f/\lambda$ are the wavenumber vectors $\boldsymbol{k}_i$ and $\boldsymbol{k}_f$, respectively. Thus, the phase difference ϕ is simply the scalar product of the difference between the wavenumber vectors $\boldsymbol{k}_i-\boldsymbol{k}_f$ and the position vector $\boldsymbol{r}$.

$$\phi=(\boldsymbol{k}_i-\boldsymbol{k}_f)\boldsymbol{r}=\boldsymbol{K}\boldsymbol{r} \tag{A.8}$$

Where the difference between the wavenumber vectors, here rewritten as $\boldsymbol{K}$, is called the **scattering vector**. By using Equation (A.8), the phase difference ϕ can easily be obtained by determining the scalar product of $\boldsymbol{K}$ and $\boldsymbol{r}$. As shown in Figure A.2, $\boldsymbol{K}$ is the vector connecting the endpoints of the incoming wavenumber vector $\boldsymbol{k}_i$ and the scattered wavenumber vector $\boldsymbol{k}_f$. Since the scattering angle is 2θ, the magnitude of $\boldsymbol{K}$ is $4\pi\sin\theta/\lambda$. After studying this section, it is hoped that the reader will appreciate the physical meaning of the variable $\sin\theta/\lambda$, so often encountered in the text, and its proportional relationship to the magnitude of the scattering vector $|\boldsymbol{K}|$.

A.4 Scattering from more than one electron

Consider the amplitude of the X-rays scattered as shown in Figure A.2. The X-rays intercept the electrons at positions P and Q from the $\boldsymbol{k}_i$ direction, and are scattered in the $\boldsymbol{k}_f$ direction. To obtain the amplitude of the scattered wave, we can simply sum the amplitudes of the waves scattered from P and Q, which differ only by phase. That is, the X-rays scattered from P and Q have the same **amplitude** (e^2/mc^2). If P is selected as the origin, the X-rays scattered at Q has phase difference by $\boldsymbol{K}\boldsymbol{r}_Q$. Accounting for this phase difference and adding the amplitudes of both scattered waves, the amplitude A of the final scattered wave is given by

$$A(\boldsymbol{K}) = \frac{e^2}{mc^2}\{1 + \exp(-i\boldsymbol{K}\boldsymbol{r}_Q)\} \tag{A.9}$$

Because of the phase difference $\boldsymbol{K}\boldsymbol{r}_Q$, the contribution of the wave scattered from Q is $\exp\{-i\boldsymbol{K}\boldsymbol{r}_Q\}$ lower than the contribution from P. The contribution $\exp\{-i\boldsymbol{K}\boldsymbol{r}_Q\}$ is called the **phase term**. Note that A is a complex number with a real and imaginary part; $A = A_r + iA_i(A^2 = A_r^2 + A_i^2)$.

Extending this argument, suppose that X-rays traveling in the $\boldsymbol{k}_i$ direction irradiate N electrons located at $\boldsymbol{r}_1, \boldsymbol{r}_2, \ldots, \boldsymbol{r}_N$. Clearly, the amplitude $A(\boldsymbol{K})$ of the scattered X-rays in the $\boldsymbol{k}_f$ direction is the sum of the phase terms of the X-rays scattered by the individual electrons.

$$A(\boldsymbol{K}) = \frac{e^2}{mc^2}\sum_{j=1}^{N}\exp(-i\boldsymbol{K}\boldsymbol{r}_j) \tag{A.10}$$

If the scatterers are continuously distributed with density $\rho(\boldsymbol{r})$, the summation in Equation (A.10) becomes an integral.

$$A(\boldsymbol{K}) = \frac{e^2}{mc^2}\int \rho(\boldsymbol{r})\exp\{-i\boldsymbol{K}\boldsymbol{r}\}d\boldsymbol{r} \tag{A.11}$$

This equation has interesting and important implications. The scattering amplitude $A(\boldsymbol{K})$ is the **Fourier integral** of the density distribution $\rho(\boldsymbol{r})$. Its square, $A^*(\boldsymbol{K})A(\boldsymbol{K})$, is proportional to the scattering intensity $I(\boldsymbol{K})$, which can be measured in a scattering experiment. Moreover, whereas the density distribution $\rho(\boldsymbol{r})$ is a function of position $\boldsymbol{r}$ in real space, the scattering intensity is a function of $\boldsymbol{K}$ in $\boldsymbol{K}$-space, which relates the scattering angle θ to the direction of the vector $\boldsymbol{K}$ {i.e., $|\boldsymbol{K}|(= \sin\theta/\lambda)$}. The $\boldsymbol{K}$- and $\boldsymbol{r}$-space are mathematically related through the Fourier transform. In fact, $\boldsymbol{K}$ has the dimensions of inverse length, as is easily suspected from $\sin\theta/\lambda$. Thus, the $\boldsymbol{K}$ space is called the **reciprocal space**.

A.5 Atomic scattering factor

When an atom is irradiated by X-rays, all of its electrons will oscillate by slightly different phases with the incident X-rays (the phase shifts will depend on the electrons' positions relative to the incoming X-rays) and will emit spherical waves of slightly different phases. The amplitude of the scattered X-rays in the direction $\boldsymbol{k}_f$ is the sum of the scattered X-rays with slightly different phases. As shown in Equation (A.11), the amplitude of the wave scattered by an atom is the Fourier integral of the density distribution of the atom ρ_{atom} $(\boldsymbol{r})$.

$$f(\boldsymbol{K}) = \int \rho_{\mathrm{atom}}(\boldsymbol{r}) \exp\{i\boldsymbol{K}\boldsymbol{r}\} d\boldsymbol{r} \tag{A.12}$$

The function $f(\boldsymbol{K})$ is called the **atomic scattering factor** or **atomic form factor**.

The electron density distribution of an atom in a crystal is spherically asymmetric because some of the electrons are shared among neighboring atoms. However, except in light elements, the proportion of asymmetrically distributed electrons is very small. Therefore, to a good approximation, we can assume that atom in a crystal is the same as isolated one in which the electrons are spherically symmetrically distributed around the nucleus. Under this approximation, Equation (A.12) is a function of r alone and reduces to

$$f(K) = 4\pi \int_0^\infty \rho(r) \frac{\sin Kr}{Kr} r^2 dr \tag{A.13}$$

where $K = 4\pi \sin\theta/\lambda$.

The calculation of the atomic scattering factor has been inspected by several authors since 1930s, by assuming the electron density distribution of atoms acceptable. In principle, if the wave function $\phi_j(\boldsymbol{r})$ is known for all of the electrons in an atom, the electron distribution function $\rho(\boldsymbol{r})$ is simply given by $\sum_j \phi_j^* \phi_j(\boldsymbol{r})$. Currently trusted values are listed in the International Tables for Crystallography **C** (1992)[9], published by the International Union of Crystallography. The reader is encouraged to refer to these. The tables report the calculated $f_j(\sin\theta/\lambda)$ values, which use the argument $\sin\theta/\lambda$ rather than K. Here, the suffix j refers to the j-th atom. As a reference, Figure A.3 plots the atomic scattering factors of several atoms (including Cs and Cl) as functions of $\sin\theta/\lambda$.

The scattering factor on the vertical axis of Figure A.3 indicates the degree of interference. The scattering factor at $\sin\theta/\lambda = 0$ is the atomic number Z of the atom. This is easily seen by substituting $\boldsymbol{K} = 0$ into Equation (A.12). The equation then reduces to $f(0) = \int \rho_{atom}(\boldsymbol{r}) d\boldsymbol{r}$, which gives the total number of electrons Z in the atom. These characteristics of the atomic scattering factor explain why the scattering intensity decreases with increasing scattering angle in Figure 1.2.

The scattering factor decreases with increasing scattering angle, indicating that fewer electrons contribute to the interference at higher scattering angles. The scattering factor of ions at zero scattering angle is larger or smaller than Z, depending on the degree of ionization. Thus, the ionization degree can be estimated from the diffraction lines at low angles. The scattering factor increases proportionally to the atomic number. Atoms with

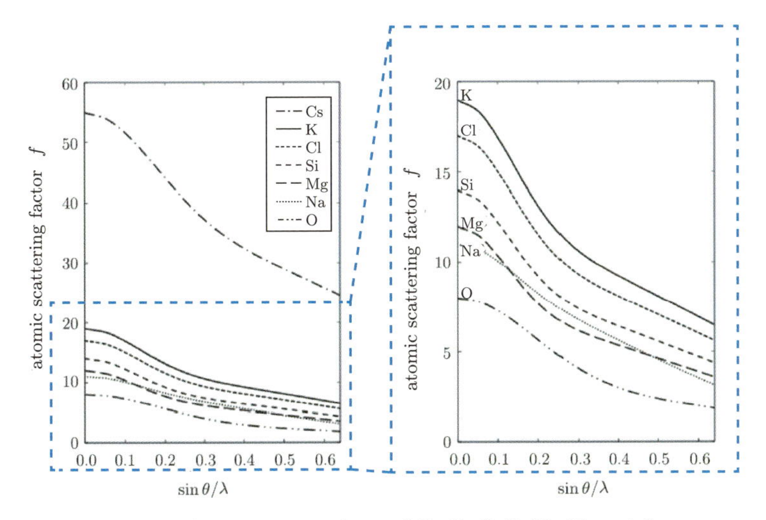

Figure A.3 Atomic scattering factor of Cs, K, Cl, Si, Mg, Na, and O as a function of scattering angle ($\sin\theta/\lambda$).

large atomic numbers are not easily distinguishable (each other) by X-ray diffraction, because the number of electrons between neighboring atoms in the periodic table changes by a relatively small amount.

We conclude with another important property of the atomic scattering factor. The above-calculated scattering factors assume a spherical electron distribution in the atom. In crystals, the electron distribution is distorted to an asymmetric shape in accordance with bonding characteristic between atoms. The distortion is especially severe in light elements, in which bound electrons constitute large ratios of the total numbers of electrons. Such cases require use of the anisotropic atomic scattering factor. Although these factors have been compiled, the reference source is quite old[12].

A.6 Crystal structure factor

Let us consider the scattering amplitude of X-rays from a unit cell. This scattering amplitude, called the **crystal structure factor** or simply **structure factor**, is the most important factor in crystal structure analysis.

As a preliminary to the structure factor, we introduced the atomic scattering factor as Equation (A.12). We now consider the scattering amplitude from an atom located at some position $\boldsymbol{r}_j$ from the origin. As illustrated in Figure A.4, the problem can easily be solved by replacing the variable $\boldsymbol{r}$ representing the charge distribution $\rho_{a,j}(\boldsymbol{r})$ of the j-th atom in Equation (A.12) by a new variable $\boldsymbol{r} - \boldsymbol{r}_j$. The scattering factor of the j-th atom is then given by the Fourier integral of the charge density $\rho_{a,j}(\boldsymbol{r} - \boldsymbol{r}_j)$ with respect to $d\boldsymbol{r}$ over all space. Note that in this calculation, the center of the charge density is displaced to $\boldsymbol{r}_j$ from the origin. Since the scattering factor of the j-th atom located at $\boldsymbol{r}_j$ can be regarded as a structure factor, it can be expressed as.

$$F_j(\boldsymbol{K}) = \int \rho_{a,j}(\boldsymbol{r} - \boldsymbol{r}_j) \exp\{i\boldsymbol{K}\boldsymbol{r}\} d\boldsymbol{r} \tag{A.14}$$

The integration variable $\boldsymbol{r}$ has been replaced by a new variable $\boldsymbol{R} = \boldsymbol{r} - \boldsymbol{r}_j$, and $d\boldsymbol{r}$ has been transformed to $d\boldsymbol{R}$. Thus, we have

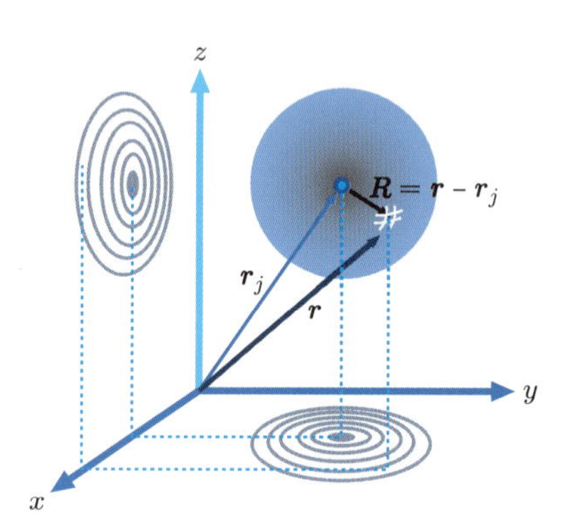

Figure A.4 Electron density distribution of a j-th atom shifted from the origin.

$$F_j(\boldsymbol{K}) = \int \rho_{a,j}(\boldsymbol{R}) \exp\{i\boldsymbol{K}(\boldsymbol{R}+\boldsymbol{r}_j)\}d\boldsymbol{R}$$
$$= \left[\int \rho_{a,j}(\boldsymbol{R}) \exp\{i\boldsymbol{K}\boldsymbol{R}\}d\boldsymbol{R}\right] \exp\{i\boldsymbol{K}\boldsymbol{r}_j\} \qquad \text{(A.14}'\text{)}$$

The term inside the square brackets [...] of Equation (A.14′) is exactly the atomic scattering factor. Accordingly, it can again be rewritten as $f_j(\boldsymbol{K})$. Thus, we have

$$F_j(\boldsymbol{K}) = f_j(\boldsymbol{K}) \exp\{i\boldsymbol{K}\boldsymbol{r}_j\} \qquad \text{(A.15)}$$

According to this equation, the amplitude of the X-rays scattered from the j-th atom is the product of the atomic scattering factor and the phase difference $\exp\{i\boldsymbol{K}\boldsymbol{r}_j\}$ introduced by the positional change from the origin to $\boldsymbol{r}_j$. Equation (A.15) is easily generalizable to several atoms located at different positions in the unit cell. Let m atoms represented by $j = 1, 2, \ldots, m$ occupy at $\boldsymbol{r}_j$ and let their atomic scattering factors be $f_j(\boldsymbol{K})$. The structure factor representing the scattering amplitude of the unit cell is obtained by multiplying the atomic scattering factors by their phase terms, and summing the products.

$$F(\boldsymbol{K}) = \sum_{j=1}^{m} f_j(\boldsymbol{K}) \exp\{i\boldsymbol{K}\boldsymbol{r}_j\} \qquad \text{(A.16)}$$

This equation gives the amplitude of the scattered wave from the unit cell in the direction given by the scattering vector $\boldsymbol{K}$. This equation is the most generalized formula for the structure factor. It should be noted that the scattering factor $F(\boldsymbol{K})$ depends not only on the magnitude $4\pi \sin\theta/\lambda$ but also on the direction of $\boldsymbol{K}$. Some readers will remember the crystallographic term "structure factor," from Chapter 2, which introduced the crystal structure factor representing the amplitude of the Bragg reflection from the lattice plane with indices h, k, and l (Equation (2.5)). We also showed that 1) Equation (2.5) describes the extinction rule and that 2) Equation (2.5) is useful for indexing the powder X-ray diffraction pattern.

To clarify the difference between the crystal structure formula in Equation (2.5) and the general structure factor in Equation (A.16), we restate both equations below.

$$F_{hkl} = \sum f_j(\sin\theta_{hkl}/\lambda)\exp\{2\pi i(hx_j + ky_j + lz_j)\} \tag{2.5}$$

$$F(\boldsymbol{K}) = \sum f_j(\boldsymbol{K})\exp\{i\boldsymbol{K}\boldsymbol{r}_j\} \tag{A.16}$$

Equation (2.5) represents the scattering amplitude by a unit cell. Because we know the scattering angles $2\theta_{hkl}$ that admit Bragg reflections in powder X-ray diffraction measurements, we do not require the scattering vector $\boldsymbol{K}$. Equation (A.16) also gives the scattering amplitude from the unit cell, but not in terms of the Bragg reflection; instead, it defines $F(\boldsymbol{K})$ as a continuous function of the scattering vector $\boldsymbol{K}$, without considering the diffraction or reflection from a crystal lattice. In the next section, we show that substituting the Bragg condition into the scattering vector $\boldsymbol{K}$ reduces Equation (A.16) to Equation (2.5).

A.7 The reciprocal lattice

The magnitude of the scattering vector $\boldsymbol{K}$ is $4\pi\sin\theta/\lambda$, and the scattering angle is 2θ. Let the scattering angle that occurs Bragg reflections from the $h\ k\ l$ lattice plane be $2\theta_{hkl}$. The condition of Bragg reflection is given by $2d_{hkl}\sin\theta_{hkl} = \lambda$. The magnitude of the scattering vector $|\boldsymbol{K}_{hkl}|$ is therefore $2\pi/d_{hkl}$. For cubic crystal, the absolute value of the scattering vector $|\boldsymbol{K}_{hkl}|$ can be obtained by

$$|\boldsymbol{K}_{hkl}| = \frac{2\pi}{d_{hkl}} = \frac{2\pi}{a_0}\sqrt{(h^2 + k^2 + l^2)} \tag{A.17}$$

where we have used Equation (1.4). On the other hand, the $\boldsymbol{K}$ vector is defined in a space given by the Fourier transform of the electron density distribution $\rho(\boldsymbol{r})$ in the unit cell, as discussed in our introduction to Equation (A.16). To represent $\boldsymbol{K}$ in such a space, we introduce three nonlinear basis vectors $\boldsymbol{a}^*, \boldsymbol{b}^*, \boldsymbol{c}^*$. $\boldsymbol{K}$ is then written as follows.

$$\boldsymbol{K} = \boldsymbol{a}^*\xi + \boldsymbol{b}^*\eta + \boldsymbol{c}^*\zeta \tag{A.18}$$

Where ξ, η, and ζ are the coordinates along the $\boldsymbol{a}^*, \boldsymbol{b}^*$, and $\boldsymbol{c}^*$-axes, respectively. These parameters are non-integer, because $\boldsymbol{K}$ is a continuous variable.

Let us consider Bragg reflection by crystal. $\boldsymbol{K}$ is now discrete, because it exists only under the Bragg condition. Here, we seek the condition for which the magnitude of $\boldsymbol{K}$ satisfies Equation (A.17), where h, k, and l are integers, and $\boldsymbol{a}^*, \boldsymbol{b}^*$, and $\boldsymbol{c}^*$ are expressed in Cartesian coordinates (also known as rectangular or orthogonal coordinates). Under the condition $|\boldsymbol{a}^*| = |\boldsymbol{b}^*| = |\boldsymbol{c}^*| = 2\pi/a_0$, $\boldsymbol{K}_{hkl}$ can be calculated as follows.

$$\boldsymbol{K}_{hkl} = \boldsymbol{a}^*h + \boldsymbol{b}^*k + \boldsymbol{c}^*l \tag{A.19}$$

A vector drawing will confirm the consistency between this equation and Equation (A.17). To determine the directions of the unit vectors $\boldsymbol{a}^*, \boldsymbol{b}^*$, and $\boldsymbol{c}^*$, we consider the Bragg reflection of the (1 0 0) plane in the crystal, with $h = 1$, $k = 0$, and $l = 0$. For Bragg reflections of the (1 0 0) plane, the scattering vector is $\boldsymbol{K}_{hkl} = \boldsymbol{a}^*$, perpendicular to the (1 0 0) plane in the crystal. Therefore, $\boldsymbol{a}^*$ is perpendicular to the unit vectors $\boldsymbol{b}$ and $\boldsymbol{c}$ in real space. The scattering vectors of the (0 1 0) and (0 0 1) planes are also perpendicular to their respective planes: under the Bragg reflection condition, $\boldsymbol{b}^*$ is perpendicular to the unit vectors $\boldsymbol{c}$ and $\boldsymbol{a}$, and $\boldsymbol{c}^*$ is perpendicular to the unit vectors $\boldsymbol{a}$ and $\boldsymbol{b}$. Therefore, the basis vectors $\boldsymbol{a}^*$, $\boldsymbol{b}^*$, and $\boldsymbol{c}^*$ in $\boldsymbol{K}$ space are related to the unit vectors $\boldsymbol{a}$, $\boldsymbol{b}$, $\boldsymbol{c}$ in real space as follows.

$$\begin{array}{lll} \boldsymbol{a}^*\boldsymbol{a} = 2\pi & \boldsymbol{b}^*\boldsymbol{a} = 0 & \boldsymbol{c}^*\boldsymbol{a} = 0 \\ \boldsymbol{a}^*\boldsymbol{b} = 0 & \boldsymbol{b}^*\boldsymbol{b} = 2\pi & \boldsymbol{c}^*\boldsymbol{b} = 0 \\ \boldsymbol{a}^*\boldsymbol{c} = 0 & \boldsymbol{b}^*\boldsymbol{c} = 0 & \boldsymbol{c}^*\boldsymbol{c} = 2\pi \end{array} \tag{A.20}$$

Here we have introduced a new lattice, whose unit cell is given by the unit vectors $\boldsymbol{a}^*$, $\boldsymbol{b}^*$, and $\boldsymbol{c}^*$ in $\boldsymbol{K}$-space. This lattice, called the **reciprocal lattice**, is closely related to the crystal lattice described by the basis vectors $\boldsymbol{a}$, $\boldsymbol{b}$, and $\boldsymbol{c}$ in **real space**. In Equation (A.19), h, k, and l are integers, so all points specified by $\boldsymbol{K}_{hkl}$ are lattice points in the reciprocal lattice. However, the reader should recognize that $\boldsymbol{K}$ is a physical quantity represented by Equation (A.18); $\boldsymbol{K}$ itself can take any non-integer value. Now suppose that X-rays are scattered by a crystallite. The reciprocal lattice of this crystallite can be constructed from all of the lattice points that allow Bragg reflections at the observation point. This concept is valid for any crystalline samples belonging to a particular crystalline system. We now relate the scattering vector to Bragg reflection in the reciprocal lattice, taking the simplest case (a cubic crystal) for convenience.

Given that the scattering vector $\boldsymbol{K}$ can be represented in terms of the reciprocal lattice, let us calculate the phase difference $\exp\{i\boldsymbol{K}\boldsymbol{r}_j\}$, the scalar product of the scattering vector (defined in reciprocal space) and the atomic position (in real space). In terms of the lattice vectors $\boldsymbol{a}$, $\boldsymbol{b}$, and $\boldsymbol{c}$ of the unit cell and the atomic coordinates $\boldsymbol{r}_j$ (x_j, y_j, z_j) of the j-th atom, $\boldsymbol{r}_j$ is given as follows.

$$\boldsymbol{r}_j = \boldsymbol{a}x_j + \boldsymbol{b}y_j + \boldsymbol{c}z_j \tag{A.21}$$

By calculating the scalar product of Equations (A.19) and (A.21) under the condition of Equation (A.20), one can easily see that $\exp\{\boldsymbol{K}_{hkl}\boldsymbol{r}_j\}$ is in agreement with the phase factor of Equation (2.5). Consequently, the reader will hopefully appreciate that under the Bragg reflection conditions, Equation (2.5) can be derived from Equation (A.16), the general formula of the structure factor.

Annotation

1) In general, X-ray scattering from electrons has commonly explained by two processes. If energy is transferred from the X-ray to the charged particle, the scattering is

inelastic. When no energy is transferred, the scattering is elastic. During inelastic scattering, energy transfer is necessarily accompanied by momentum transfer. This book focuses on the elastic scattering, but an example of the inelastic scattering is given below.

The inelastic X-ray scattering by electrons is called **Compton scattering**. As mentioned many times throughout this text, X-rays are electromagnetic waves, but they also behave as particles (photons). When a photon with energy $h\nu$ collides with a stationary electron, some of its energy is transferred to the electron, reducing the frequency of the photon. This frequency change characterizes Compton scattering. The energy loss is evaluated by the angle of scattering.

The inverse scattering process is called **Inverse Compton scattering**. For instance, the visible light which is the photon collide with an accelerated electron with high energy. The visible light with relatively low energy photons deprives some of the kinetic energy which electron had; the low energy photon gains energy by the scattering and changes to a high energy photon. The author has heard such an attempt to get a strong X-ray source using this phenomenon.

Appendix B Geometry of crystal lattice

We can state without exaggeration that most of today's industrial materials are crystalline substances. The structures of these materials can be identified and evaluated by X-ray diffraction. A fundamental knowledge of crystal structure is essential to understand X-ray diffraction data. Thus, this Appendix provides a basic discussion of crystal lattices.

B.1 Space lattice and unit cell

Crystals are formed by the orderly and periodic arrangements of atoms, ions, or molecules, which can be large or small. These arrangements are called **space lattices**. A periodic structure is a structure consisting of repeated basic units, and is said to demonstrate **translational symmetry**. The basic unit is called a **unit cell**. The unit cell of one-dimensional periodic structure can be determined unambiguously, but the unit cell of two- and three-dimensional periodic structure can take various configurations. As a guideline, the unit cell should be as simple as possible, which means that the crystal lattice of the cell can be described by few parameters. Once the arrangement of atoms or molecules in the unit cell is determined, the structure of that crystal is known to a certain extent. Currently, researchers are searching beyond the molecular and atomic configurations, examining the distributions of the bonded electrons and the anisotropic thermal oscillation of atoms. Such information is often acquired in crystal structural studies.

If one considers the three-dimensional periodicity of crystals, one will realize that the unit cell must be a parallelepiped. A standard unit cell is

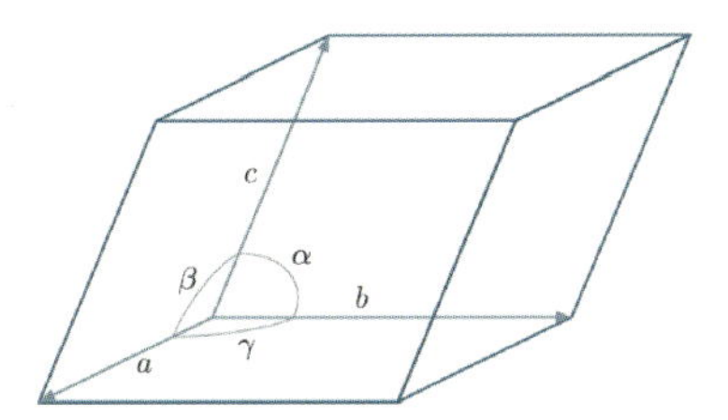

Figure B.1 Unit cell and its parameters.

shown in Figure B.1. A mathematical description of this unit cell requires six parameters: the three lengths of the rhomboid sides, $\boldsymbol{a}$, $\boldsymbol{b}$, and $\boldsymbol{c}$, and the three angles formed by the rhomboids, α, β, and γ. The unit cell may also be expressed by three basis vectors, $\boldsymbol{a}$, $\boldsymbol{b}$, and $\boldsymbol{c}$. In terms of the basis vectors, any point on the crystal lattice can be expressed as follows.

$$\boldsymbol{r}_{mnp} = m\boldsymbol{a} + n\boldsymbol{b} + p\boldsymbol{c} \tag{B.1}$$

In Equation (B.1), m, n, and p are integers, but since crystals are finite, it is reasonable to assume that these integers are also finite.

B.2 Crystal system

Figure B.1 illustrates a standard unit cell. Repetition of this unit cell forms a three-dimensional lattice whose points exhibit certain symmetric properties. In considering these properties, we assume an infinite lattice and consider the symmetry elements of the selected lattice points.

If we connect any two lattice points and examine the symmetry properties about the line connecting those points, we can identify the following symmetry elements.

1) $\boldsymbol{n}$-fold rotational axis ($\boldsymbol{n} = \mathbf{1}, \mathbf{2}, \mathbf{3}, \mathbf{4},$ and $\mathbf{6}$) : Symbolized by $\boldsymbol{n}$ (360/$\boldsymbol{n}$ degree rotation)

When rotated $360/n$ ($n = 1, 2, 3, 4,$ and 6) degree around a certain axis, the three-dimensional lattice completely overlaps another lattice, which is then indistinguishable from the original lattice.

2) Reflection plane: Symbolized by $\boldsymbol{m}$ (reflection)

A lattice plane with reflection symmetry behaves like a mirror. When a lattice plane is regared as a mirror, the image of the upper half lattice on the mirror overlap with the lower half lattice, and both the lattices becomes indistinguishable.

3) Inversion centre: Symbolized by $\overline{\mathbf{1}}$ (inversion center)

Select a lattice point. When the vector from the lattice point to a selected origin is $\boldsymbol{r}$, a newly configured lattice point is located at the $-\boldsymbol{r}$ position. If the newly configured lattice points are indistinguishable from the original lattice points, it has an inversion center.

4) Rotoinversion axis: Symbolized by $\overline{\mathbf{2}}$ $(= m)$, $\overline{\mathbf{3}}$ $(\equiv \mathbf{3} \times \overline{\mathbf{1}})$, $\overline{\mathbf{4}}$, and $\overline{\mathbf{6}}$ $(\equiv \mathbf{3}/\boldsymbol{m})$ (total of 5)[9]

The lattice returns to its original position after combined rotation and inversion operations.

Symmetry operations performed on combinations of the above symmetry elements are also symmetry elements. Mathematically, these symmetry operations form 32 **point groups** in total.

Space lattices with these 32 symmetry elements are classified by seven types, based on the shape of the unit cell. These are called the **7 crystal systems**.

We now describe the methodology and terminology for assigning symmetry elements. First, let us examine the main axis in a given unit cell, which is not necessarily the $\boldsymbol{a}$-axis. The symmetry element of this axis is written

Table B.1 7 crystal system.

Crystal system	Parameters		Number of parameter (Laue classes)	
Triclinic	a≠b≠c	$\alpha \neq \beta \neq \gamma \neq 90$ deg.	6	$(\bar{1})$
Monoclinic	a≠b≠c	$\alpha = \gamma = 90$ deg., $\beta \neq 90$ deg.	4	(2/m)
Orthorhombic	a≠b≠c	$\alpha = \beta = \gamma = 90$ deg.	3	(mmm)
Trigonal (rhombohedral axes)	a=b=c	$\alpha = \beta = \gamma \neq 90$ deg.	2	$(\bar{3}, \bar{3}m)$
Tetragonal	a=b≠c	$\alpha = \beta = \gamma = 90$ deg.	2	(4/m, 4/mmm)
Hexagonal	a=b≠c	$\alpha = \beta = 90$ deg., $\gamma = 120$ deg.	2	(6/m, 6/mmm)
Cubic	a=b=c	$\alpha = \beta = \gamma = 90$ deg	1	$(m\bar{3}, m\bar{3}m)$

first. Thereafter, we identify and write down the symmetry elements of the second axis intersecting the main axis. The notation "/" immediately after the corresponding symmetry element denotes that the two axes cross at right angles. Thus, $\mathbf{2}/\boldsymbol{m}$ signifies a two-fold rotational main axis with a reflection plane at right angles (/) to that axis. Meanwhile, $\mathbf{4}/\boldsymbol{mmm}$ denotes a four-fold rotational main axis, a reflection plane at right angles to the four-fold rotation axis $(/\boldsymbol{m})$, and two additional reflection planes $(\boldsymbol{mm})$. The reader is advised to check these symmetry elements in the unit cell in Figure B.2.

When translational symmetry is added to the symmetry elements of the space lattice and the **symmetry of the atomic configuration** in the unit cell is also considered, we have 230 possible combinations of symmetry elements. Therefore, three-dimensional atomic configurations can be classified into 230 types (called **space groups**). Any crystal can be assigned to one of these space groups. However, when X-rays are irradiated onto crystals, the three-dimensional symmetry of the resulting diffraction spots reveals only a group of lattice points with a center of symmetry. These elements, called **Laue classes**, show the number only 11. Although the three-dimensional lattices formed by the diffracted spots must not be confused with crystal lattices, both exhibit three-dimensional space lattices. Crystal lattices can also be expressed in terms of the Laue groups, as illustrated in Figure B.2.

B.3 Bravais lattice

Each of 7 crystal systems listed in Table B.1 includes a single lattice point in the unit cell. These lattices are called **primitive lattices**. Refer

Crystal system	Simple	Base-centered	Body-centered	Face-centered
Triclinic				
Monoclinic				
Orthorhombic				
Trigonal				
Tetragonal				
Hexagonal				
Cubic				

Figure B.2 Bravais lattice.

their visualized images shown in the first column of the Figure B.2. After slightly displacing the duplicate from the original lattice position, you obtain a compound lattice called a **complex lattice**. To maintain the same symmetry as the original lattice, more lattices of the same type must be combined into the complex lattice. However, the number of combined lattices is limited. From the seven original and seven additional lattices, one can create 14 complex lattices. This type of lattice was first discovered by Auguste Bravais (1811–1863) and is hence called the **Bravais lattice**, as shown in Figure B.2.

By changing the way of taking the lattice vectors of the complex lattices, a primitive lattice can be created. In such a case, the crystallographic axis changes from that shown in Figure B.2, and the lattice belongs to a different crystal system. In addition, the atomic coordinates become non-intuitive. As an example, Figure B.3 shows a primitive lattice to the face-centered cubic lattice. This crystal system is triclinic rather than cubic. It is important to realize the possible ambiguity of selecting a crystal system. To avoid confusion, the axis should never be selected haphazardly. Ideally,

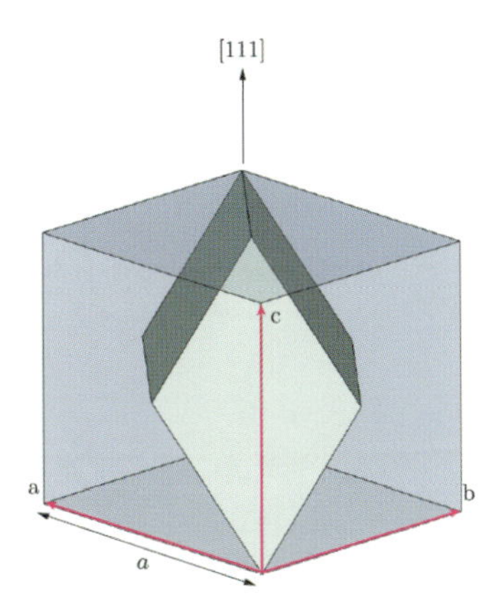

Figure B.3 Relationship between FCC unit cell and its primitive unit cell.

the main axis should have a high degree of symmetry. If discovered an unknown crystal, it is recommended to determine the appropriate crystal system from specialized references.

B.4 Lattice plane and Miller indices

In a three-dimensional lattice, three lattice points not located on a single line can be connected to describe a plane. Such a plane is called a **lattice-plane** (net plane of the lattice) or a crystallographic plane. Therefore, a three-dimensional lattice can be regarded as an infinite number of parallel planes with constant **interplanar spacing** (or lattice-plane spacing, d-spacing).

Since X-ray diffraction is closely associated with the lattice planes, these lattice planes need to be named to prevent confusion. To this end, we identify a specific plane in a lattice by three integers, h, k, and l, which are called the **Miller indices**. The lattice plane assignment is based on the following rules.

As shown in Figure B.4, we describe a plane intersecting unit vector $\boldsymbol{a}$ at $1/h$, unit vector $\boldsymbol{b}$ at $1/k$, and unit vector $\boldsymbol{c}$ at $1/l$. When this plane is extended, it always crosses the lattice points of the $\boldsymbol{a}$-, $\boldsymbol{b}$-, and $\boldsymbol{c}$-axes. You may confirm easily this by a two-dimensional lattice. The group of planes parallel to this lattice plane can also be expressed by the three integers h, k, and l. Since this concept was derived from William Hallowes Miller

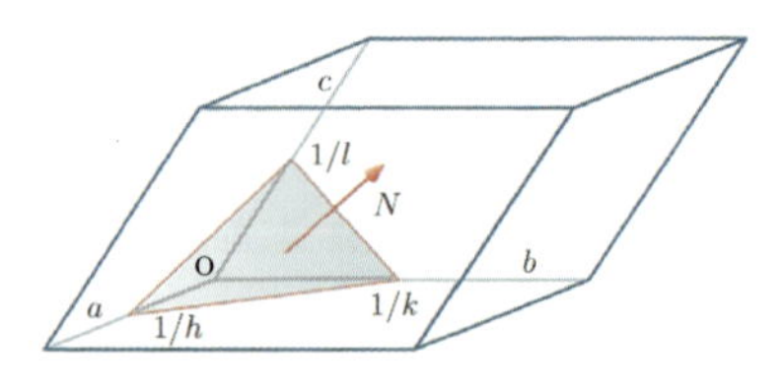

Figure B.4 (h k l) lattice plane.

(1801–1880), these integers are called **Miller indices** (or **plane indices**). Although different texts offer different explanations, all of them explain the same idea.

Select any three points A, B, and C in a space lattice that are integral multiples of the unit vectors $\boldsymbol{a}$, $\boldsymbol{b}$, and $\boldsymbol{c}$, and label them h', k', and l', respectively. We then seek three integers, h, k, and l that maintain the same ratio as the inverses of h', k', and l' (namely, $1/h'$, $1/k'$, and $1/l'$). In other words, h, k, and l are prime integers satisfying the following conditions: $h = m/h'$, $k = m/k'$, and $l = m/l'$. These integers h, k, and l are called the Miller indices (or plane indices) of the above-mentioned lattice plane.

The interplanar spacing of the parallel lattice planes with indices, h, k, and l, is the distance from the origin to the plane. Thus, the above expression is convenient for evaluating the interplanar spacing. A single lattice plane or a plane group parallel to that plane is written as plane $h\ k\ l$ or $(h\ k\ l)$. The **International Union of Crystallography** (IUCr) has established various rules for labeling planes. The parentheses symbolize the lattice plane itself; the direction perpendicular to the plane is indicated in square brackets $[h\ k\ l]$. A space lattice also contains a large number of equivalent lattice planes, all expressed as $\{h\ k\ l\}$ with orientations indicated by $\langle h\ k\ l \rangle$.

Let us examine some specific examples. A lattice plane parallel to the crystallographic axis is said to intersect that axis at a distance far from the origin (actually, at infinity). Consider a lattice plane intersecting the $\boldsymbol{a}$-axis at lattice point a, and the $\boldsymbol{b}$- and $\boldsymbol{c}$-axes at infinity. The inverses of the intersecting lattice points are written as $1/1$, $1/\infty$, and $1/\infty$, with integer ratios of 1, 0, and 0, respectively. Thus, this plane is a $(1\ 0\ 0)$ lattice plane. Next, consider a lattice plane intersecting the $\boldsymbol{a}$-axis at lattice point $-a$, and the $\boldsymbol{b}$- and $\boldsymbol{c}$-axes at infinity. This plane is $(-1\ 0\ 0)$, written as $(\bar{1}\ 0\ 0)$ in crystallography notation, and can be understood as a plane viewed from the back side of the plane $(1\ 0\ 0)$.

There are cases that crystal of which looking from the back side is different to that from the front side. A layered crystal consisting of dual atomic layers A and B can be shown as this example. The configuration along the direction prepedicular to the layer is given as A–B$\cdots$A–B$\cdots$A–B$\cdots$. This kind of crystal is useful for understanding a difference of the viewing side of the crystallographic plane. A real example is given in Section C2.2 of Appendix C.

Extending this discussion further, consider a plane that intersects the $\boldsymbol{a}$-axis midway between two lattice points, and intersects the $\boldsymbol{b}$- and $\boldsymbol{c}$-axes at infinity. This plane, designated the (2 0 0) plane, is parallel to the (1 0 0) plane and indicates planes with half the interplanar spacing of the (1 0 0) plane. It can easily be seen that an (h 0 0) plane is a plane with an interplanar spacing of $1/h$. Therefore, the Miller indices specify the interplanar spacing and the orientation of the lattice plane. Figure 1.5 (b) shows the typical lattice planes in a cubic crystal system. Readers should familiarize themselves with these planes.

B.5 Interplanar spacing

Another important parameter of space lattices is the interplanar spacing. As mentioned in the previous section, a space lattice consists of numerous lattice planes. A lattice plane is characterized by its interplanar spacing and the normal direction to the plane. Here, focusing on the orientation and interplanar spacing of the lattice planes given by the indices h, k, and l, we discuss how these quantities are determined.

First, consider the unit vector $\boldsymbol{n}$ along the direction normal to a lattice plane. The interplanar spacing d_{hkl} is the scalar product of $\boldsymbol{n}$ and $\boldsymbol{a}/h$, $\boldsymbol{b}/k$, or $\boldsymbol{c}/l$. The relationship is expressed as follows.

$$\left(\frac{\boldsymbol{a}\cdot\boldsymbol{n}}{h}\right)=\left(\frac{\boldsymbol{b}\cdot\boldsymbol{n}}{k}\right)=\left(\frac{\boldsymbol{c}\cdot\boldsymbol{n}}{l}\right)=|d_{hkl}| \tag{B.2}$$

Table B.2 Relationship between crystal system and the interplaner spacing .

Crystal system	Interplaner spacing *hkl*
Triclinic	$(1/d_{hkl})^2 = \{(h^2/a^2)\sin^2\alpha + (k^2/b^2)\sin^2\beta + (l^2/c^2)\sin^2\gamma$ $+(2hk/ab)(\cos\alpha \cos\beta - \cos\gamma)$ $+(2kl/bc)(\cos\beta \cos\gamma - \cos\alpha)$ $+(2lh/ca)(\cos\gamma \cos\alpha - \cos\beta)\}$ $/(1 - \cos^2\alpha - \cos^2\beta - \cos^2\gamma + 2\cos\alpha \cos\beta \cos\gamma)$
Monoclinic	$(1/d_{hkl})^2 = \{(h^2/a^2) + (k^2/b^2) - (2hk\cos\gamma/ab)\}/\sin^2\gamma + (l^2/c^2)$ $(1/d_{hkl})^2 = \{(h^2/a^2) + (l^2/c^2) - (2hl\cos\beta/ac)\}/\sin^2\beta + (k^2/b^2)$
Orthorhombic	$(1/d_{hkl})^2 = (h^2/a^2) + (k^2/b^2) + (l^2/c^2)$
Trigonal	$(1/d_{hkl})^2 = \{(1/a^2) + (h^2 + k^2 + l^2)\sin^2\alpha$ $+2(hk + kl + lh)(\cos^2\alpha - \cos\alpha)\}$ $\times(1 + 2\cos^3\alpha - 3\cos^2\alpha)^{-1}$
Tetragonal	$(1/d_{hkl})^2 = (h^2 + k^2)/a^2 + l^2/c^2$
Hexagonal	$(1/d_{hkl})^2 = (4/3a^2)(h^2 + hk + k^2) + (l^2/c^2)$
Cubic	$(1/d_{hkl})^2 = (1/a)^2(h^2 + k^2 + l^2)$

In cubic/orthorhombic crystal (in which the $\boldsymbol{a}$-, $\boldsymbol{b}$-, and $\boldsymbol{c}$-axes intersect at right angles) and hexagonal crystal systems, the relationship between d_{hkl} and the lattice parameters can easily be derived from geometric considerations. In a cubic crystal system, the relationship becomes $(1/d_{hkl})^2 = (1/a)^2(h^2 + k^2 + l^2)$. In other crystal systems such as monoclinic, triclinic, and rhombohedral systems, the derivation is more complicated, but is simplified by the concept of reciprocal lattice, which was introduced in Appendix A (Basic principles of X-ray scattering and diffraction). Table B.2 lists the d_{hkl} values associated with given $h\ k\ l$ in all crystal systems.

Appendix C

Cohesive force and crystal structure

The crystallite is formed by aggregation of atoms and molecules, and the material is texture of the crystallites. If we classified the material by their cohesive force, it is divided into five, metallic bond, ionic bond, covalent bond, hydrogen bond and coupling between the molecules combined by the Van der Waals force. Based on this classification, we pick up some typical crystals representing each cohesive force. In that case, we may consider that there is a certain relationship between the crystal structure and their bonding force. However, even crystals bonded by different cohesion will notice that there are a number of those having the same crystal structure. Besides, the most crystals are stable in a certain temperature range but change into different crystal structure when the temperature exceeds the range. Such a change of crystal structure is called a phase transition. Paying an attention also to this phase transition, showing the facts in relationship between the cohesive force and the crystal structure is the focus of Appendix C.

C.1 Crystal structures of metallic materials

The agglomeration mechanism of metallic crystals is formed in a way that outer-shell electrons which have no restrictions imposed to specific ions are shared within a periodic arrangement of ions comprising the nuclei and inner-shell electrons. Although it said that ion was carrying out periodic arrangement, it does not necessarily need to be periodic, because of existence of amorphous state. This is indicating that the metal is formed by a balance of the long-range Coulomb attractive force between the array ions and free electrons to the short-range repulsive force between the nearest neighbor ions. Therefore, the crystal structure can be said to be an array itself of the ions.

C.1.1 Three basic structures

The crystal structure of the metal is summarized in Table C.1. Most metals show one of the three simple structures, the body centered cubic (BCC), the face centered cubic (FCC), and **hexagonal close packed structure** (HCP), although exclusively Mn, Ga and In are of different crystal structure. The table includes the hydrogen and the elements of inert gas besides metals. In metals, the most of elements which belong to the same column are of the same structure. It will be related to the energy of conduction electron.

Three basic structures are shown in Figure C.1. The number of atom, the atomic position, and the number of nearest neighbor atoms in the unit cell are summarized in Table C.2. As easy to see from Figure C.1 the number of the nearest neighbor atoms is 8 for the BCC structure, but 12 for the FCC and the HCP structure. Although the FCC structure seems to be different from the HCP structure at a glance, both structures are strongly related to each other. That is why both the structures have 12 nearest neighbor atoms. When viewed from the [1 1 1] direction, the FCC lattice appears as a 2-dimensional atomic plane comprised of hexagonal lattices formed by arranged atomic spheres of the same size. When atoms are positioned on the recessed areas formed on the atomic plane, a new atomic plane is created. This is called the **stacking layer structure**. The (1 1 1) plane of the FCC structure is the same arrangement as the (0 0 1) plane of the HCP structure. However, both structures are slightly different in the way of stacking.

Table C.1 Crystal structure of elements. Each column shows the chemical symbol, the crystal structure such as BCC, FCC, HCP, the lattice parameter a (c is written in the case of HCP), and the axis ratio c/a.

Ia	IIa	III a	IV a	V a	VI a	VII a	⟵	VIII a	⟶	b	II b	III b	IV b		VIII b
H HCP 3.75 6.12 c/a: 1.632															He HCP 3.75 5.83 c/a: 1.633
Li BCC 3.491	Be HCP 2.27 3.59 c/a: 1.581														Ne FCC 4.46
Na BCC 4.225	Mg HCP 3.21 5.21 c/a: 1.623											Al FCC 4.05			Ar FCC 5.31
K BCC 5.225	Ca FCC 5.58	Sc HCP 3.31 5.27 c/a: 1.592	Ti HCP 2.95 4.68 c/a: 1.586	V BCC 3.03	Cr BCC 2.88	Mn cubic complicated	Fe BCC 2.87	Co HCP 2.51 4.07 c/a: 1.621	Ni FCC 3.51	Cu FCC 3.61	Zn HCP 2.66 4.95 c/a: 1.861	Ga complicated			Kr FCC 5.64
Rb BCC 5.585	Sr FCC 6.08	Y HCP 3.65 5.73 c/a: 1.569	Zr HCP 3.23 5.15 c/a: 1.590	Nb BCC 3.30	Mo BCC 3.15	Tc HCP 2.74 4.40 c/a: 1.606	Ru HCP 2.71 4.28 c/a: 1.579	Rh FCC 3.80	Pd FCC 3.39	Ag FCC 4.09	Gd HCP 2.98 5.62 c/a: 1.886	In tetragonal			Xe FCC 6.13
Cs BCC 6.045	Ba BCC 5.02	La HCP 3.77 complicated	Hf HCP 3.19 5.05 c/a: 1.583	Ta BCC 3.30	W BCC 3.16	Re HCP 2.76 4.46 c/a: 1.616	Os HCP 2.74 4.32 c/a: 1.576	Ir FCC 3.84	Pt FCC 3.92	Au FCC 4.08	Hg	Tl HCP 3.46 5.52 c/a: 1.595	Pb FCC 4.95		Rn

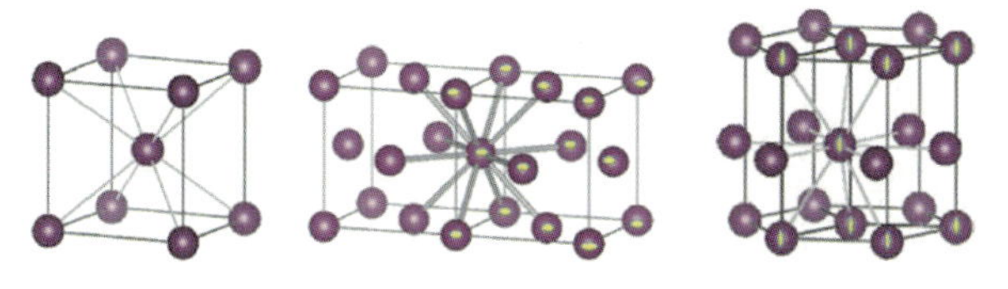

(a) Body-centered cubic (BCC) (b) Face-centered cubic (FCC) (c) Hexagonal close-packed structure (HCP)

Figure C.1 Typical crystal structures of metal. (b) shows two adjacent unit cells and (c) shows three adjacent units; the nearest neighbor atoms are drawn by light grey colored thick lines.

Table C.2 Crystal structure and atomic coordinate.

structure	number of atom in unit cell	atomic coordinate		number of nearest neighbor atoms
BCC	2	0, 0, 0	1/2, 1/2, 1/2	8
FCC	4	0, 0, 0	1/2, 1/2, 0	12
		0, 1/2, 1/2	1/2, 0, 1/2	
HCP	2	0, 0, 0	2/3, 1/3, 1/2	12

The Figure C.2 (a) and (a′) show the FCC and the HCP structure, respectively. The Figure C.2 (b) and (b′) are the projection of the (1 1 1) plane of the FCC structure, and that of the (0 0 1) plane of the HCP structure, respectively. If the atomic position of the bottom plane is assigned to be A-site, we see that there are two types of the hollow sites, each formed by three atoms. They are the position (blue circle) of the center of gravity of the equilateral minimun yellow triangle and that (red circle) of an inverted minimun yellow triangle. Suppose the positions of blue and red circle be the B-site and C-site, respectively. We see that the FCC structure is formed by stacking of the 2-dimensional hexagonal lattice on A-site, B-site, and C-site repeatedly in sequence. This is called ABC stacking. By changing the view, if the first atomic lattice plane is formed by A-site, the next atomic lattice plane can be formed by putting the atoms on the hollow B-site of the first atomic lattice plane. If repeated exactly this process again, the third atomic lattice layer is formed by putting the atoms on the hollow B-site of the second atomic plane. This is the FCC stacking structure.

Structure of the HCP is also a stacking structure. In this case, after upper atomic layer takes B-site of the basic A-site layer, the next atomic layer goes back again to the A-site, without taking the three sites A, B and C. Thus, the stacking consists of two sites arranged repeatedly in the order of A, B, A, B and so on. This is called an AB stacking structure. The [1 1 1]

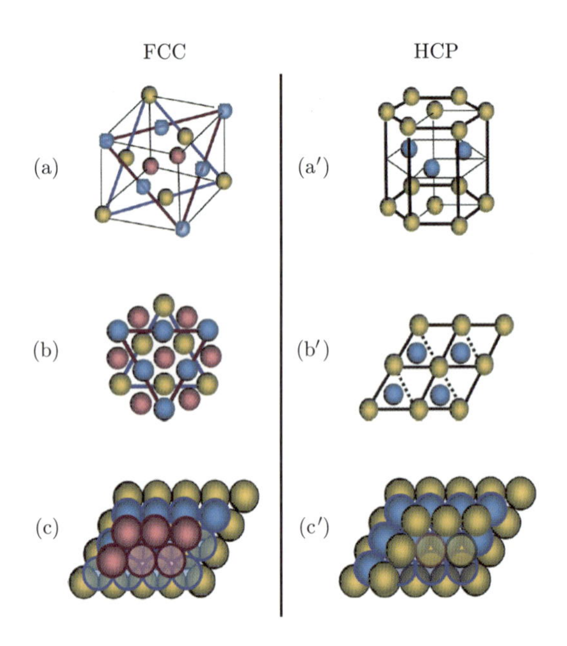

Figure C.2 Difference of stacking layer structure in FCC and HCP1).
(a) and (a′) are a unit cell of the FCC and the HCP, respectively. (b) and (b′) are the projection view along [1 1 1] and [0 0 1] direction, respectively.
(c) and (c′) shows three dimensional stacking layer structure of both structures.

direction of the FCC structure and the [0 0 1] direction (c-axis direction) of the HCP structure can be said to be the characteristic crystallographic axes. The Figure C.2 (c) and (c′) are solid figures, representing the difference between the FCC and the HCP stacking structure.

Let's examine the differences in the diffraction patterns obtained from three types of crystal structure of metals. Since we have studied the diffraction patterns of α-Fe featuring a BCC crystal structure and Al featuring an FCC crystal structure in Chapter 1, we discuss here the diffraction pattern of Co, which shows an HCP crystal structure. The Figure C.3 shows the diffraction pattern for Co powder sample of HCP structure obtained by using MiniFlex 300/600. A remarkable feature of this diffraction pattern is in a fact that the three diffraction peaks are observed closely together at low angle side. These are indexed as the (1 0 0), (0 0 2), and (1 0 1) diffraction peaks, respectively. If familiarize with diffraction pattern, it is

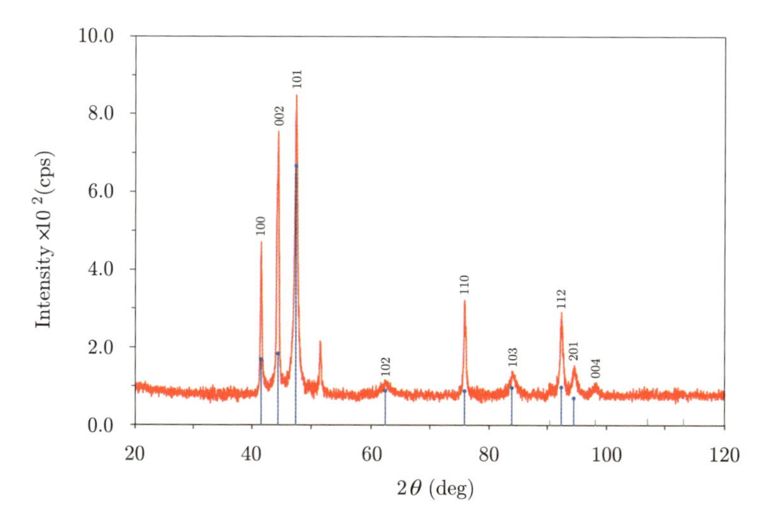

Figure C.3 X-ray diffraction pattern of Co powder sample measured by CuKα. The peak around $2\theta = 51.5$ degree denotes that it includes small quantity of FCC phase.

possible to distinguish above three crystal structures by just looking at the patterns.

By the way, if we assume that the HCP structure is created by stacking of two-dimensional atomic layers formed by atoms of the ideal sphere, the axial ratio c/a of HCP is determined only by the geometry of the structure. In the case, it is denoted by $c/a = (8/3)^{1/2} = 1.633$. However, the metal of the ideal value is seen only in He, as shown in the Table C.1. The only other metals close to this ideal are Mg with 1.623, and Co with 1.622. All others are far from the ideal value. Does this mean that atoms are not assumed as spherical? This is an interesting question, and a number of other basic questions will be raised in relation to this issue.

C.1.2 Phase transition

In the previous section, only the typical crystal structures of simple metals have been considered. One more thing in the Table C.1, Fe has been represented by BCC structure[2)] but this structure is stable until up to 1184 K (911°C) when heated up from room temperature. Fe having such a BCC structure in this range of temperature is called the **ferrite**. Beyond this temperature, the crystal changes to the FCC structure that known as **austenite** phase. When further raised the temperature up to 1392 K, it changes again to the BCC structure called **δ-ferrite** and takes this structure until to the melting point 1537 K.

Change of a crystal structure to another structure at a certain temperature as shown above is called the **phase transition** in the solid phase. When the transition occurs from solid to liquid, it is well known fact that the latent heat is required for the melting. Similarly, when a structural phase change happens at a certain temperature, a certain amount of heat energy is required also for the migration of constituent atoms or molecules in the rearrangement of structure.

As already mentioned above, Fe shows two phase transitions, but only two stable crystal structures exist from room temperature to melting point. There is one more point to be mentioned here. Pure metals are rarely used for practical applications. Metals used on a daily basis are in fact mixtures of several metals, which are called alloys. In general, the alloy changes also the crystal structure by mixing ratio and temperature. As a result, the characteristics of the alloy such as its mechanical and electrical properties also change. To understand such characteristics, the possible changes in crystal structure have to be studied.

Again using iron to illustrate our point, we note that pure iron is rather soft. However, iron mixed with a small amount (up to 1.7 %) of carbon creates steel, which has high hardness. Stainless steel is an alloy of iron mixed with Cr and Ni, and is well known because it does not oxidize, unlike pure iron. The crystal structure of iron alloys also depends on its mixing percentage. The mixing percentage determines whether an iron alloy takes the form of a nonmagnetic austenite or a magnetic ferrite. Thus, each alloy has specific structure and temperature range with stable structure.

C.1.3 Disordered structure

As the simplest example of the alloy, **brass** (CuZn) of a binary alloy of copper and zinc is shown. Pure Cu has the FCC structure, whereas pure Zn has the HCP structure. The FCC lattice of Cu does not change by mixing up to about 35 atomic % of Zn atoms; this solid solution is called the α-phase. As in the example of CuZn, if the diameters of the two atoms are close together, the range of composition of the solid solution has been widely observed. The Zn atoms in the α-phase are intermingling in FCC lattice of the matrix phase of Cu. This is called **disordered structure**.

In contrast, when the Zn reaches about 50 atomic %, it would be difficult to predict which structure does it take; FCC structure of Cu or HCP structure of Zn. A funny thing happens because the structure is BCC, which is called β-phase, not FCC and HCP. The β-phase is the disordered state forming complete solid solution between Cu and Zn at high temperature over about 360°C. This transition temperature shifts from 454°C up to 468°C depending on the narrow range of composition ratio of 45 – 50 atomic % Zn in Cu. Below the transition temperature, disordered β-phase changes into an **ordered state** called β'-phase. The structure of β'-phase is of the CsCl type so that Cu atom occupied (0 0 0) position and Zn is in (1/2, 1/2, 1/2), or vice versa, which has been already mentioned in the Chapter 1. All the Bragg reflections are observed in the CsCl structure, but some of the reflections disappeared in the BCC structure due to the extinction rule. The reflection peaks observed only in the ordered phase are called the **super-lattice reflection**. If you observe the temperature dependence of integrated intensity of such a super-lattice reflection from a low temperature up to over the transition temperature, you can confirm the transition temperature by disappearance, or the extinction, of the super-lattice reflection. It is attributed to change from the ordered structure to the disordered structure of Cu and Zn by the structural phase transition, as shown in the Figure C.4. Usually the disordered structure do not appear so suddenly; the super-lattice reflection peak decreases gradually even at a temperature below the transition temperature. The phase tran-

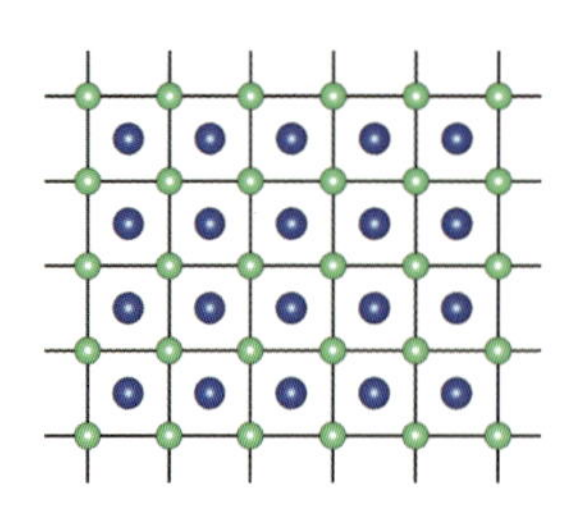

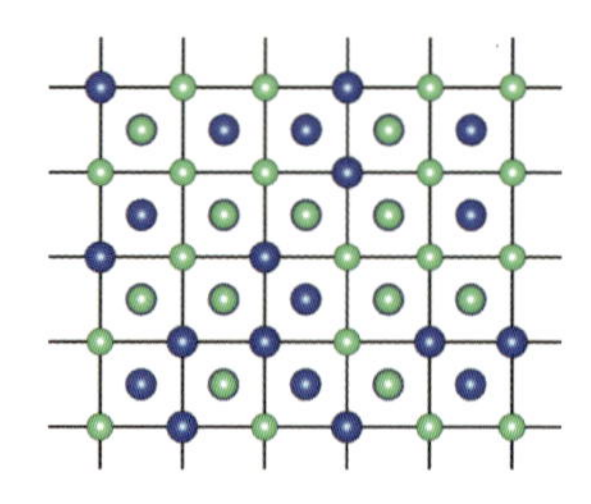

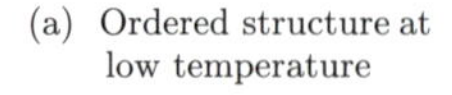

(a) Ordered structure at low temperature

(b) Disordered structure at high temperature above 360℃

Figure C.4 Two dimensional schematic drawings of the structures for ordered and disordered state of CuZn.

sition therefore emerges gradually. The structure accompanied with such a phase transition is called the **short-range order structure**, implying the existence of locally ordered structure.

C.1.4 Stability of crystal structure

In Appendix C, we have closely examined the simple metal crystal to learn the relationship between cohesive force and crystal structure. We found that the metal has three kinds of structures. Among the metals, we selected the Fe as a familiar material and studied the property and the structure up to the melting temperature. As a result, Fe shows the structural phase transition at an elevated temperature. In addition, Fe changes to another material with different mechanical property, if included tiny amount of carbon. We have understood that the structure of metals changes depending on not only the composition of alloy but also its temperature. It has been noticed that cohesive force has a great influence on the crystal structure of metal and alloy. Generally crystal structure is not however settled only on the cohesive force.

Through the example it has been shown that the crystal structure changes into another one at a certain temperature, if the temperature of a crystal was increased. This fact is understood as the same phenomenon as seen in the melting of a material. With the increasing temperature, atoms or molecules begin to move against their cohesive force. With the building up of the atomic or molecular motion, the structure itself is getting to be unstable. The atoms and molecules could not form their own structure and they would freely move, because it would be stable than the crystal. Such a phenomenon would happen very locally at the beginning and extend finally it to whole crystal. This is known as the melting.

A basic situation of the phase transition is very similar to this melting. After the phase transition, a different structure is more stable than the original structure. Because the thermal movement of the atoms or molecules would be easier on the new crystal structure than on previous structure. In the thermodynamics, it says that the state having a low Gibbs free energy is more stable than the other. This is based on the idea that the stability of a phase is not dependent only on the internal energy U and it is controlled by its Gibbs free energy G, which should be lowest compared with that of any other phases. G is defined by $G = U + PV - TS$, where S is the entropy of the system being the physical quantity to express the degree of freedom, T is the absolute temperature. P and V are the pressure and the volume, respectively. If PV of the solid state could be regarded to be almost unchanged at absolute zero temperature, the Gibbs free energy will be a function of U only.

In the crystal structure obtained through the analysis of a given crystalline material by the X-ray diffraction method, we understand 1) there is a temperature range in which the crystal structure obtained is stable. In addition, 2) the crystal structure changes when there are any contaminants. During your experiment, please pay attention to above facts.

C.2 Crystal structures of ionic crystals

A NaCl is a typical ionic crystal in which Na^+ cation and Cl^- anion are alternately arranged by Coulomb force. In general, ionic crystal is composed of positively charged ions and negatively charged ions by arranging themselves with Coulomb force. The elements belonging to Ia of the periodic table such as Li, Na, K, Rb, and Cs become easily positive ions by emitting one outer shell electron. On the other hand, the nonmetallic atoms in the group Ⅶb of the periodic table such as F, Cl, Br, I,... become negative ions relatively easily by acquiring one electron. The charge density distribution of both the ions are spherically symmetric, since the outer shell electrons are of a closed packed structure of the $(sp)^8$ type. The bonding energy of the ionic crystal is bigger than that of metal, due to Coulomb force which influences up to long range. The ionic crystal is classified by its structure. We pick up here following five kinds of ionic crystals: CsCl, NaCl, ZnS, wurtzite (this is a polymorph structure of ZnS) and perovskite-type. The polymorph (polymorphic form) denotes the crystal having a different crystal structure, although the chemical formula is the same.

C.2.1 CsCl-type and NaCl-type

The Figure C.5 shows the typical crystal structure of CsCl-type crystal. The unit cell is shown in (a) and a partial sketch of some extended region of crystal structure is shown in (b). As seen from the unit cell, Cs^+ ion is located at the body center position (1/2, 1/2, 1/2), while Cl^- ion is located at the corner of cubic lattice (0, 0, 0). The crystal is electrically neutral because unit cell contains one CsCl molecule. You may think that the charge balance is not kept because the CsCl molecule with a dipole is directed towards the direction of $[\bar{1}\ \bar{1}\ \bar{1}]$ in the unit cell. However, one cation in the body-centered position is in fact balanced by the contribution of 1/8 anion at 8 corners (the number of anions in the unit cell is thus $8 \times 1/8 = 1$), which balances the charge within unit cell. To confirm the validity of this approach, we can extend this counting method to an aggregation of large number of unit cells. The whole will be $CsCl_8$ with a negative charge. Then, let us imagine a little big ion group by incorporated with six of the second nearest neighbor Cs ion. These Cs ions are repelled from the Cs ions of the center because of the same type ion, but those six second nearest neighbors are attracted to the eight of the Cl ions. Thus, a whole a negative big ion cluster of Cs_7Cl_8 is formed. In this way, when we alternately add the number of ions to this cluster, a large ion cluster will gradually be grown up. The central part of the cluster is getting to become stable increasingly with the increase of its size. Its reason is that Coulomb force, which is long-range force acting between many ions,

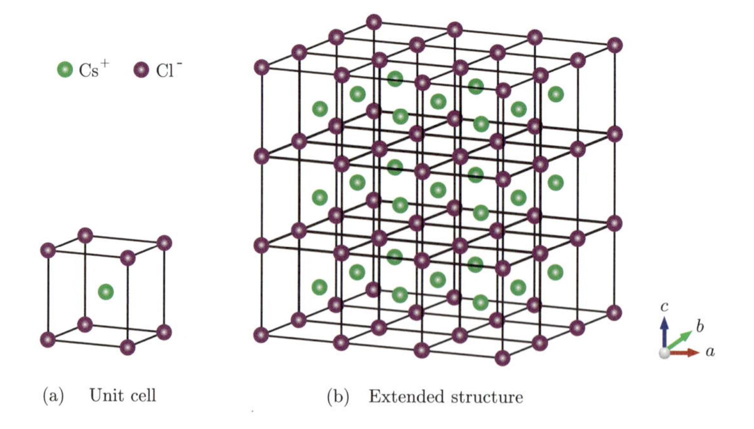

(a) Unit cell (b) Extended structure

Figure C.5. The unit cell and extended structure of CsCl type crystal.

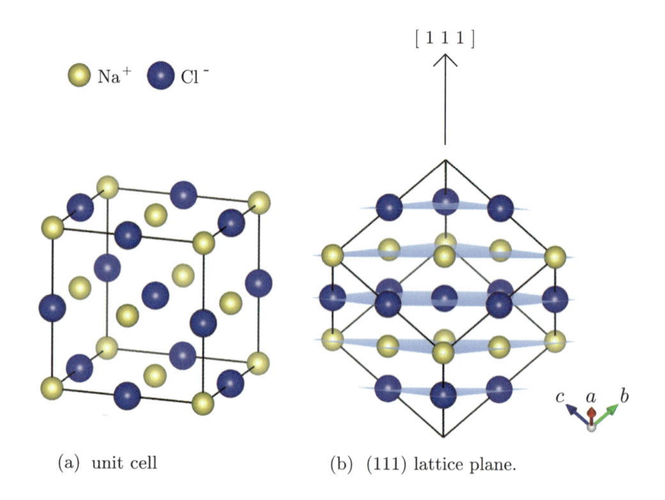

(a) unit cell (b) (111) lattice plane.

Figure C.6 Structure of NaCl type crystal.

Table C.4 Compounds with NaCl type structure. The number written in lower part indicates the lattice parameter in Å unit[17]. The number inside parentheses is the number of ICDD.

LiF	NaCl	KCl	KBr
4.027 (4-857)	5.640 (5-628)	6.295 (74-9685)	6.601 (36-1471)
AgBr	MgO	MnO	PbS
5.775 (79-149)	4.211 (76-2583)	4.217 (76-6596)	5.931 (78-1901)

becomes attractive and takes a balance with the repulsive force between the core electrons of ions.

The Figure C.6 shows the unit cell of NaCl-type structure. It is the cubic structure and 4 NaCl molecules exist in the unit cell. The structure is considered to be formed by two face centered cubic lattices of cations and anions which are separated by 1/2 along the [0 0 1] direction. Thus, Na cation's locations are at the face center position, (0, 0, 0), (1/2, 0, 1/2), (0, 1/2, 1/2), (1/2, 1/2, 0), and Cl anions are at (0, 0, 1/2), (1/2, 0, 1), (0, 1/2, 1) and (1/2, 1/2, 1/2). In this case, we count the contribution of the ion located at a face center position (0, 1/2, 1/2) to be 1/2, and that of the ion located at a corner (0, 0, 1/2) to be 1/4. In this NaCl structure, one of the ions is always surrounded by the 6 different kinds of ions as the nearest neighbors and by the 12 same kind of ions as the second nearest neighbors. The crystals showing this structure are listed in the Table C.4.

The Figure C.7 is the diffraction pattern of the NaCl measured by Mini-Flex 300/600. The structure is the FCC structure and characteristic diffraction peaks, which are similar to that of Al, are observed. Two reflection peaks of (1 1 1) and (2 0 0) are seen in the low angle side. The (2 2 0) reflection is observed in a slightly higher angle from the (2 0 0) reflection. In comparison with the intensity of diffraction peaks of Al, the difference comes out as a fact that the intensity of the (1 1 1) reflection in NaCl is prominently weaker than that of the (2 0 0) reflection. The reason can be

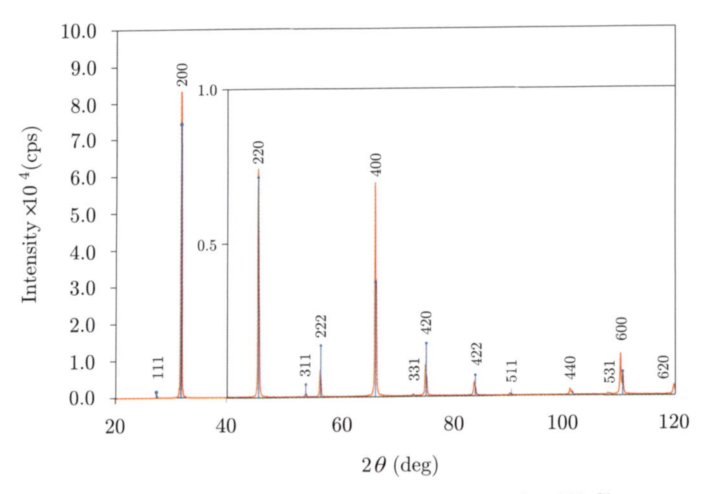

Figure C.7 X-ray diffraction pattern of powdered NaCl measured by CuKα.

understood by looking out the ionic stacking of the (1 1 1) plane in NaCl. The stacking order is in such a way that the lattice plane formed of Cl ions gets into just between the lattice planes formed of Na ions. Hence, the wave scattered by the lattice plane formed of Na ions has a difference by a half phase from the wave scattered from the lattice plane of Cl ions. Two waves will interfere destructively, leading the intensity reflected a weak. The KCl crystal is exactly the same structure as that of NaCl. However, the K-ion and the Cl-ion have the same valence electrons, hence a strong destructive interference occurs between the two waves in (1 1 1) reflection, making it a weak reflection that is scarcely observed. Therefore, the intensity in the diffraction pattern is changed by the difference of the total number of electrons of the ion, even if the structure is the same as that of NaCl.

C.2.2 Zinc blende and Wurtzite

The Figure C.8 (a) shows the **zinc blende structure** formed by Zn^{2+} and S^{2-} which belong to group Ⅱb and Ⅵb, respectively. The blue sphere

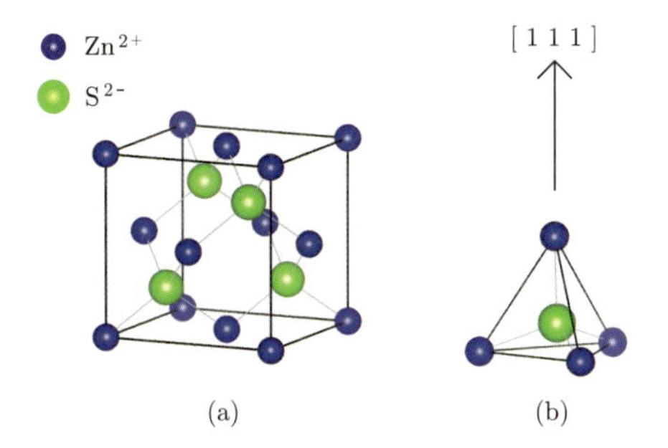

(a) Arrangement of ions in the unit cell

(b) Each ion is surrounded by four ions which are of another kind and form a tetrahedron.

Figure C.8 ZnS type structure

and the green sphere show the Zn^{2+} and the S^{2-} ion, respectively. The structure is the cubic crystal. The FCC lattice of Zn^{2+} and the FCC lattice of S^{2-} are shifted each other by 1/4 along the body-centered direction. The 4-ZnS molecules are in the unit cell. The ion position is as follows.

The position of Zn^{2+} : (0, 0, 0), (1/2, 0, 1/2), (0, 1/2, 1/2), (1/2, 1/2, 0)

The position of S^{2-} : (1/4, 1/4, 1/4), (3/4, 1/4, 3/4), (1/4, 3/4, 3/4), (3/4, 3/4, 1/4)

The nearest neighbor of the ion is 4; it is less number in comparison with CsCl and NaCl. The characteristic of the structure is in the **tetrahedral structure** surrounded by ions. In addition, the number of a second nearest neighbor is not so many as 8 of the same kind of ions.

Let us look at the tetrahedral arrangement of ions of the $(\bar{1}\ 1\ 1)$ lattice plane by setting the $[\bar{1}\ 1\ 1]$ axis as z-axis, as shown in the Figure C.9 (a). If paid attention to sulfur ions described by green ball, you can see the structure in the stacking of ABCABC $\cdots$ along the $[\bar{1}\ 1\ 1]$ direction, since it is the FCC structure originally. When we viewed the structure by including

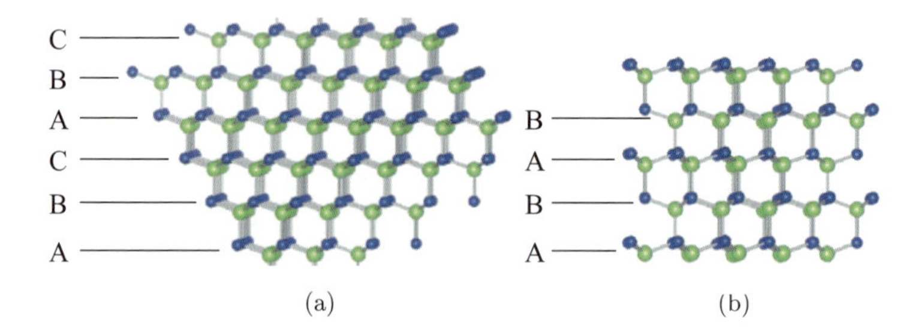

Figure C.9 The structural difference of (a) ZnS type and (b) wurtzite type structures. The electric double layer is formed by Zn^{2+} (blue circle) and S^{2-} (green circle). In (a) , those double layers form the stacking structures of ABCABC···type, which is seen along the [111] direction of the ZnS structure. On the other hand, those layer of (b) form a little different stacking structure of ABAB··· type, which is seen along the c-axis of the wurtzite structure.

Zn ion, it is understood to be the stacking structure of ABCABC ······ of which element is formed by an electric double layer of Zn^{2+} and S^{2-}. If there exists a stacking of such an electric double layer of ABCABC······, there should be stacking of the electric double layer of ABAB ·······. The polymorph materials with hexagonal strncture, not only ZnS, show the ABAB ······ electric double layer. They are listed in Table C.5. The structure is called the **hexagonal zinc-sulfide structure** or the **wurtzite structure**. Interesting point is that the c/a is close to the ideal value of 1.633 for the materials of the wurtzite structure except SiC.

Although we have mentioned until now only the ionic crystals formed by the two groups Ⅱb and Ⅵb, there are also a lot of material formed by other group. As shown in the Table C.5, there is a combination of Ⅰb – Ⅶb group as well as Ⅲb – Ⅴb group represented by GaN which is focused as the blue light emitting diode. This is telling that various combinations such as $(Ga_xAl_{1-x})N$ are possible artificially. Hence, the compound semiconductor crystals are attracting attention as functional materials. When the ionic valence increases, it is difficult to say whether ionization energy of any atom

Table C.5 Crystals of ZnS type and wurtzite type.
The lattice parameter (a for cubic, a and c for hexagonal) and the number of ICDD are shown. The unit of the lattice parameter is Å.

Crystal	ZnS type (cubic)	Wurtzite type (hexagonal) a c	Group
SiC	4.358 (75-254)	3.081 15.1 (75-8314)	IVb - VIb (covalent bonding)
AlF	4.32	-	IIIb - VIIb
CuF	4.264 (71-3775)	-	Ib - VIIb
CuCl	5.420 (82-2117)	-	Ib - VIIb
AgI	6.067 (78-749)	4.594 7.51 (83-581)	Ib - VIIb
α AgI	5.063 (74-2433)	-	Ib - VIIb
ZuO	4.270 (73-8589)	3.25 5.21 (89-1397)	IIb - VIb
ZnS	5.417 (71-5975)	3.821 6.25 (36-1450)	IIb - VIb
ZnSe	5.674 (71-5977)	3.996 6.62 (89-2940)	IIb - VIb
ZnTe	6.106 (71-5963)	4.310 7.09 (19-1482)	IIb - VIb
CdS	5.830 (89-440)	4.140 6.71 (74-9663)	IIb - VIb
CdSe	6.05 (65-2891)	4.26 6.94 (75-5680)	IIb - VIb
InP	5.869 (32-452)	-	IIIb - Vb
InAs	6.058 (15-869)	-	IIIb - Vb
InSb	6.478 (6-208)	-	IIIb - Vb
GaN	4.503 (52-791)	3.189 5.18 (89-8624)	IIIb - Vb
GaP	5.451 (32-397)	3.759 6.174 (80-2)	IIIb - Vb
GaAs	5.653 (14-150)	3.912 6.441 (80-3)	IIIb - Vb

decreases. It has been known a case in which cohesive bonding between atoms is stronger rather than ionic bonding. That is the case of Ⅳb−Ⅵb combination group such as SiC which is classified into covalent crystal rather than an ionic crystal.

Here, there is a notable characteristic, as seen in the zinc blende structure of Figure C.8 (a) and (b). The structure has no mirror symmetry between the top and bottom of a plane perpendicular to the [1 1 1] direction. The distance between the Zn plane and the S plane is extremely narrow. The crystal having such a structure is called the polar crystal. In the wurtzite structure, you can find a similar structure along the z-axis.

As for a reference of the difference in the diffraction pattern between the zinc blende structure and the wurtzite structure, both the diffraction patterns are shown in the Figure C.10 (a) and (b). In the diffraction pattern of ZnS type structure, the (1 1 1) and the (2 0 0) reflections are observed at low angle, and also the (2 2 0) reflection shows at a slightly high angle. Those three reflections are of the characteristic of FCC structure. In comparison with the intensity of diffraction peak of Al, the intensity of (2 0 0) is prominently weaker than that of (1 1 1). The reason can be seen in the Figure C.8. When we look at the stacking along the [2 0 0] direction, the two lattice planes of the Zn and the S ions exist alternately. When we look at along the [1 1 1] direction, the binary layers of ZnS maintain the sane lattice spacing of the (1 1 1) plane. On the other hand, the diffraction pattern of ZnO of the wurtzite type in the Figure C.9 (b) is seen with relatively high intensity in the three diffraction peaks of the (1 0 0) (0 0 2) (1 0 1), which is the characteristic of hexagonal crystal.

C.2.3 CaF_2 type crystal

As another example of the ionic crystal, there is CaF_2 type crystal. Until now, we have examined the ionic crystals formed by two kinds of ions. That is the ionic charge is the same, but the crystal is consisted of two

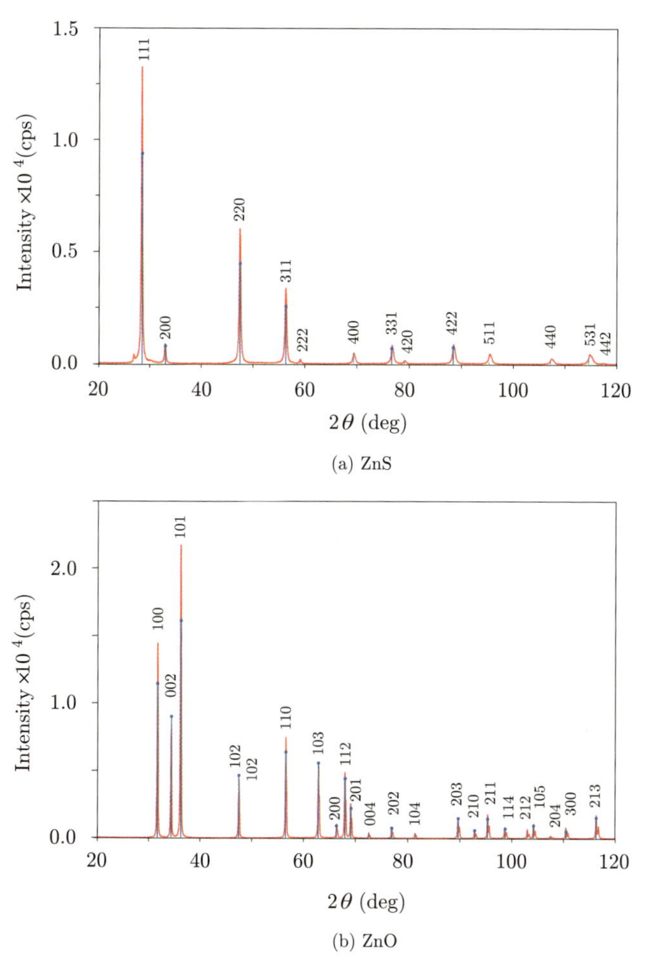

Figure C.10 X-rays diffraction pattern for (a) powdered ZnS and (b) ZnO of wurtzite type structure.

different kinds of ions A^+ and B^-. Here, we consider a case of the crystal type formed by divalent cation and two monovalent anions. It maintains electrical neutrality. Figure C.11 (a) shows the unit cell of CaF_2. The Ca^{2+} ion is in the FCC position, and F^- ion is placed at (1/4, 1/4, 1/4) and (1/4, 3/4, 1/4). The position of the F^- ion conforms also with the FCC arrangement when the position is designated as the origin. Crystals with

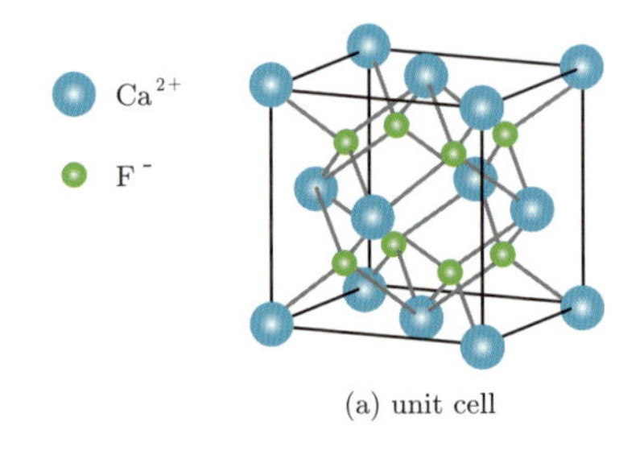

(a) unit cell

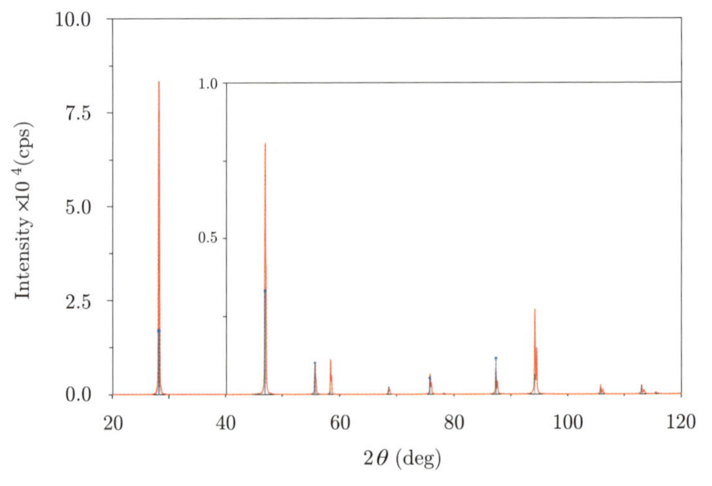

(b) X-ray diffraction pattern of powdered CaF_2, measured by MiniFlex 300/600. The red line is the observed line and the blue line is the intensity obtained by the ICDD search.

Figure C.11. Unit cell and X-ray diffraction pattern of CaF_2.

Table C.6 Ionic crystals of CaF_2 type.

Crystal	Lattice parameter	Number of ICDD	Group
CaF_2	5.463	(35 - 816)	IIa - VIIb
SrF_2	5.800	(88 - 2294)	
$SrCl_2$	6.977	(6 - 537)	
BaF_2	6.200	(4 - 452)	
CdF_2	5.390	(23 - 864)	IIb - VIIb
PbF_2	5.931	(77 - 1865)	IVb - VIIb
CsO_2	6.620	(65 - 9025)	Ia - VIb
PrO_2	5.392	(24 - 1006)	IIIa - VIb
ThO_2	5.604	(71 - 6407)	
ZrO_2	5.100	(71 - 4810)	IVa - VIb
Li_2O	4.613	(76 - 9262)	Ia - VIb
Li_2S	5.719	(77 - 2145)	
Na_2S	6.539	(23 - 441)	
Cu_2S	5.564	(53 - 522)	Ib - VIb
Cu_2Se	5.760	(65 - 2982)	

such a $CaFe_2$ type are listed in the Table C.6. As shown in the Table C.6, it is generally composed by the divalent alkali ion and the monovalent halogen ion. In addition, there are lots of cases composed by the divalent anion and the monovalent cation, as shown in the example of Cu_2S (Cu^+, S^{2-}); the anion and the cation belong to the group VIb and the group Ib, respectively, in the periodic table. The structure maintains the neutrality. It can be shown by the laminated structure along the [1 1 1] direction, The (1 1 1) layer is formed by the layer of divalent ions sandwiched by the two layers of monovalent anions.

Using CuKα line, the X-ray diffraction pattern of $CaFe_2$ powder sample is shown in the Figure C.11 (b). The red lines are the observations. The blue lines with dot are the intensities of the Bragg reflections at the scattering angles given by ICDD search. You may see that the observation corresponds well with the database of ICDD.

C.2.4 Rutile-type crystal

Although the chemical formula AX_2 is the same as the CaF_2, there is TiO_2 (Titanium dioxide), showing completely different crystal structure. The structure of the TiO_2 has been known to exist three kinds of polymorphs of the **rutile**, the **anatase** and the **brookite**. Among them, the rutile has a high melting point of 1830 °C and shows a high refractivity. The rutile has been utilized as jewelry and referred to as the **titania**. It is also used in the cosmetics due to a high infrared absorption, bactericidal effect, and high stabilization with a low price. Including the rutile, TiO_2 can be said to be the material that is considered to have still a great deal of applications.

The brookite belongs to the orthorhombic crystal system. The anatase and the rutile are of the tetragonal system although the structures are different. We here show the structure of the rutile in Figure C.12. The diffraction pattern has been shown in the Chapter 7. The rutile is the tetragonal crystal that the tetragonality (c/a) is smaller than 1. At the room temperature, the lattice parameters are $a = 4.539$ Å and $c = 2.959$ Å. In the Figure C.12, small blue spheres are assigned to the Ti ions and large

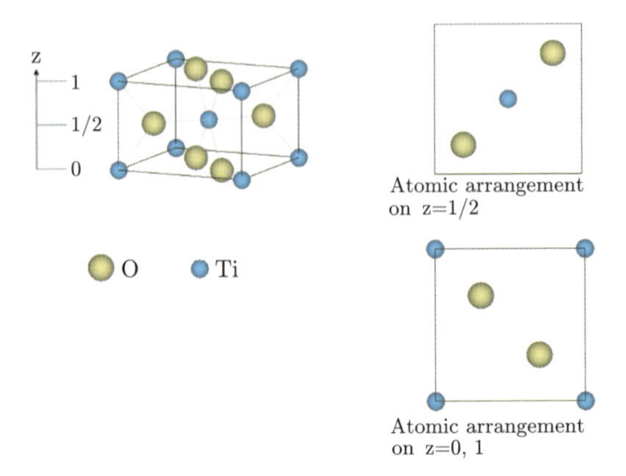

Figure C.12 Crystal structure of rutile.

yellow spheres to the O ions. The right figure shows the atom position in the plane of $z = 0$ or 1 and $z = 1/2$, for showing the atom position conveniently. The Ti atoms are arranged so that the 6 oxygen atoms form the octahedral. On the other hand, the oxygen atom is located in the center of mass of an isosceles triangle of the Ti ions. The rutile structure is also shown in the fluoride material including Mg, Ni, Co, Fe, Zn, and Mn, and the oxide material including Mn, V, Ru, Os, Nb, Sn, Ob, and Fe. The diffraction patterns of the rutile and the anatase have been described in the Section 3.3, so they are omitted here.

C.2.5 Perovskite-type crystal

As one of the ionic crystals of complex type, there is a **Perovskite structure**[15]. It is the cubic crystal and is briefly written with ABX_3 or RMX_3. The representative crystals are $CaTiO_3$ and $BaTiO_3$ and so on. The ionic arrangement in the unit cell is shown in Figure C.13 (a). The divalent cation of R^{2+} is located on the corner and the tetravalent cation of M^{4+} is on the body centered position of the cubic unit cell. In addition to these ions, the divalent anions of X^{2-} are taken the face centered positions to keep electrically the neutrality against the two cations of R^{2+} and M^{4+}. From a different viewpoint, this structure is formed by cations of R^{2+} and a cluster of octahedron made of six anions of X^{2-} and a cation of M^{4+}. The structure indicates that R^{2+} ion is positioned at the vacancy of the framework made of MX_6-octahedron.

Figure C.13 (b) shows the X-ray diffraction pattern of powdered $CaTiO_3$ measured by CuKα line. The structure is the cubic crystal. Since the extinct reflection does not exist, all the Bragg reflections are observed. In the X-ray scattering by this material, it should be understood as that the contribution of O ion is small, because O ion is of very low electron density compared with those of the Ca and Ti ion. On the other hand, the atomic number of Ca and Ti are 20 and 22, respectively. The scattering from

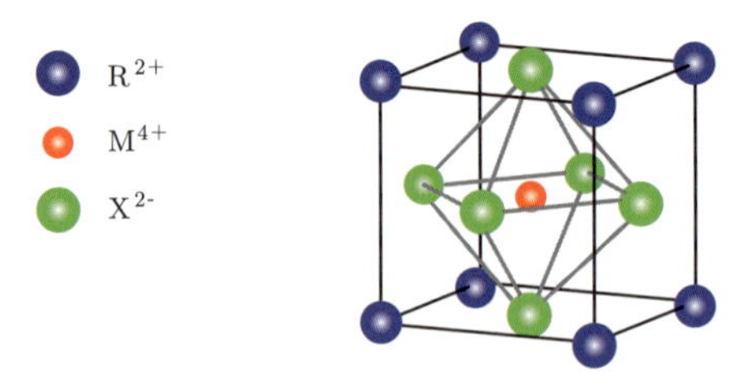

(a) RMX_3 perovskite structure.

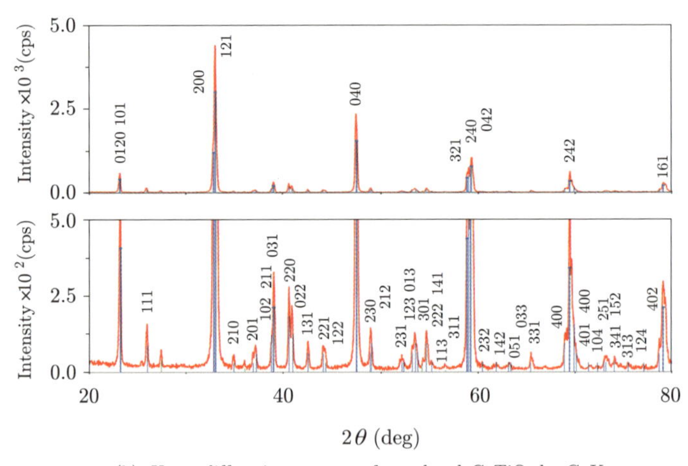

(b) X-ray diffraction pattern of powdered $CaTiO_3$ by CuKα.

Figure C.13 Perovskite structure and X-ray diffraction pattern of powdered $CaTiO_3$.

those two ions appears two of extreme cases: one is enhancing and the other is subtracting the scatterings from each scatterer. It is dependent on the reflections, since this is strongly related with the position of the BCC lattice. Thus, intensity modulation may be anticipated if based on the extinction rule appeared in the BCC lattice. Furthermore, the indexing of the diffraction pattern is not so difficult. The result is shown in the

Figure C.13 (b), in which unknown peak was found to be approximately 27 degree of 2θ. Nevertheless, this peak could be easily identified to be from TiO_2 based on the database of ICDD. Since an unknown reflection could be assigned to be TiO_2, we can say the indexing to all the reflections observed is completed.

This perovskite structure has a center of symmetry and is an insulator, having electrically neutral. The shape of MX_6 cluster is, however, easily deformed by a different size of R^{2+} and M^{4+} ions. Along with the deformation of the shape, the cubic structure changes to the tetragonal, orthorhombic, and rhombohedral structure, as seen such an example in $BaTiO_3$. Along with such a structural change, crystals significantly change their dielectric properties. Some crystals show **ferroelectricity**[14], and some others indicate antiferroelectricity. When external pressure is added to, this kind of crystals is polarized. Reversely, application of an electric field makes crystal deformed, that is called the **Piezoelectric effect**[14]. The ferroelectricity is the state of crystal in which unit cells polarized are lined up, and the anti-ferroelectricity in which unit cells polarized are alternatively lined up. These crystals show a piezoelectric effect and are widely used.

In addition, the material shows the characteristic phase transition with dielectric property change depending on the temperature. For instance, as temperature decrease from high temperature, $BaTiO_3$ undergoes a transition from paraelectric cubic crystal to ferroelectric tetragonal crystal at 120 °C, accompanying with a slight change in the lattice parameters of a-axis and c-axis. With further decrease of the temperature, it changes to orthorhombic structure around the temperature of 0 °C, and to rhombohedral structure at the temperature of −80 °C. In the $PbZrO_3$, the structural change happens, accompany with the change of dielectric property from paraelectric to antiferroelectirc at the temperature of 230 °C.

There is a phase transition from a ferroelectric to another ferroelectric phase, in which the direction of the polarization changes from one of the crystallographic axis to another. Although the phase transition is a slight deformation of crystal lattice, a bigger change of crystal symmetry is induced. In addition, $SrTiO_3$ and $LaAlO_3$ show a super-lattice transition with a change of the orientation of octahedral MX_6, without change of the external shape. Although it has mentioned that octahedron of MX_6 is easily deformed, the unit cell is not deformed in $SrTiO_3$ and $LaAlO_3$. Instead, as shown in Figure C.14 (a), MX_6 frame can rotates around the crystal axis of [0 0 1]. If described more in detail, the MX_6 octahedral frame rotates a little bit by keeping consistency with the same octahedrons which are in left and right, back and forth and up and down, as shown in the figure. Such a tiny rotation of the MX_6 induces, however, rearrangement

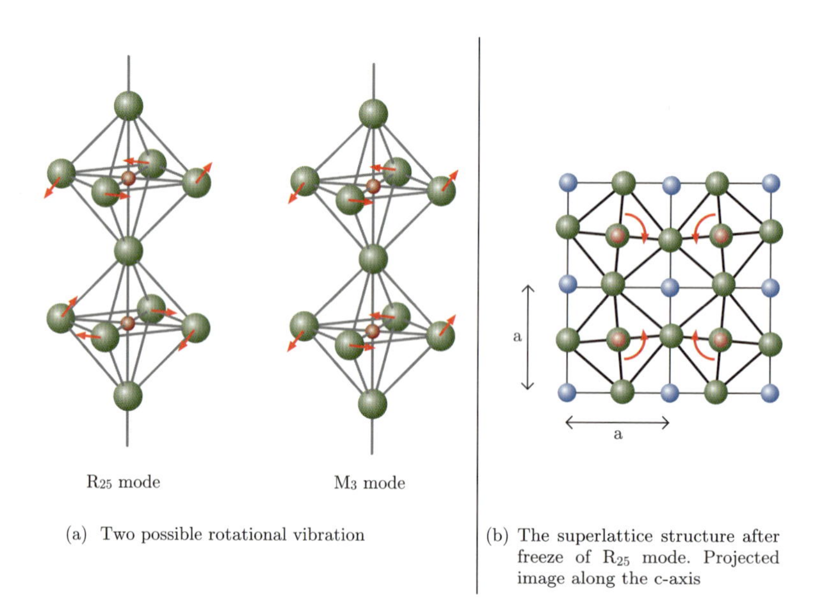

(a) Two possible rotational vibration

(b) The superlattice structure after freeze of R_{25} mode. Projected image along the c-axis

Figure C.14 Super-lattice structure of $SrTiO_3.TiO_6$ cluster tends to cause rotational vibration around a crystal axis. The structure is known to change into the super-lattice structure after freezing up of the rotational vibration.

of ions, indicating an appearance of a super-lattice structure. This is the phase transition from a simple cubic perovskite structure to a super-lattice of large unit cell. In Figure C.14 (a), the left figure shows that upper and lower octahedrons reversely rotate around the axis of [0 0 1]. The right figure shows the rotation in the same direction. Figure C.14 (b) show a consistency among the unit cells along the [0 0 1] direction. Such phase transitions are called the **structural phase transition**. A possible phase transition observed in several perovskite crystals are listed in Table C.7[15].

Here, the basic structure and the main phase transition have been shown as an example. It might be possible to produce such a new material as $(R_{1-x}{}^{a}, R_{x}{}^{b})(M_{1-y}{}^{a}\ M_{y}{}^{b}\ X_3)$, when it is formed by a mixture of more than two kinds of crystals having their own properties. Because it seems to

Table C.7 Perovskite type crystals and structural phase transition[15],[16].

Crystal	Lattice parameter	Phase transition temperature (℃)	
$BaTiO_3$	a=3.994 c=4.03 (5.626)	130, 5, -75	ferroelectric transition
$KNbO_3$	3.996 4.06 (71-945)	435, 225, -10	transformation of MX_6 structure
$PbTiO_3$	3.902 4.150 (78-298)	490	
$KTiO_3$	※	< -273	
$NaNbO_3$	3.927 3.934 (74-2455)	643, 572, 520, 480, 365	anti-ferroelectric transition
$PbZrO_3$	※	233	transformation of MX_6 structure
$PbHfO_3$	※	215	
$SrTiO_3$	5.511 7.796 (74-2455)	-168 -87, -181	phase transition to superlattice structure
$KMnF_3$	5.74 8.22 (82-1399)		due to rotation of MX_6 structure
$LaAlO_3$	※	535	
$CsPbCl_3$	5.584 5.62 (18-366)	47, 42, 37	
$CsPbiBr_3$	※	131, 89	

※ refer to [16].

be particularly easy to mixture the ceramics in a form of powder. By paying an attention to these points, development of a novel mixture material with specific properties seems to have been progressing.

C.3 Crystal structures of covalent crystals

The diamond[4] formed by carbon atoms show a typical crystal structure bounded by **covalent bonding**. Not only the diamond of C atom but also all the elements, Si, Ge, and Sn, which belong to group IVb of the periodic table, form the diamond type structure. The atoms are bonded by the covalent bonding (or homo-polar bonding). Of the bonds between atoms shown symbolically is shown in Figure C.15 (a), which is often called the Lewis structure. C, Si, and Ge have 4-outer-shell electrons but need another 4 electrons to form a stable inner-shell. The atom borrows one electron each from 4 adjacent atoms, and holds consistency with itself. On the contrary, the atom provides own 4-outer-shell electrons to 4 adjacency atoms. That is to say, each atom in the crystal keeps a stable closed-shell by sharing two electrons with each of 4 adjacency atoms. This is the covalent bonding and homo-polar bonding in another way; the crystal bonded by such a cohesive force is called the **covalent crystal**.

C.3.1 Diamond

A unit cell structure of the diamond is shown in the Figure C.15 (b). It is the structure occupied by one kind of atom of IVb group in the periodic table without making distinction between cation and anion in the zinc-blende structure. It can be said that two FCC structures are shifted by 1/4, 1/4, 1/4 each other. The 8 atoms are occupied by the following position in

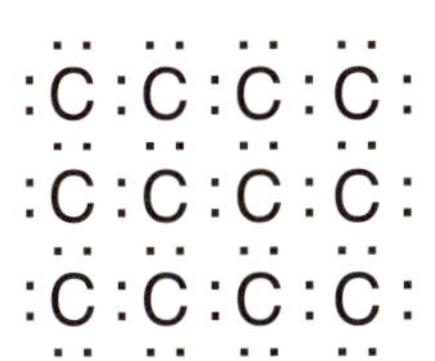

(a) Lewis structure for covalent bonding

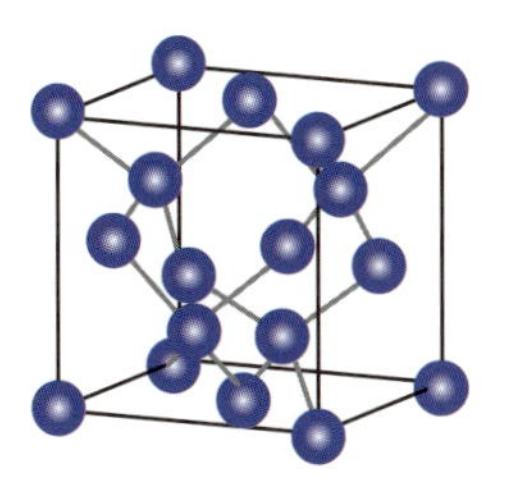

(b) Unit cell of diamond structure

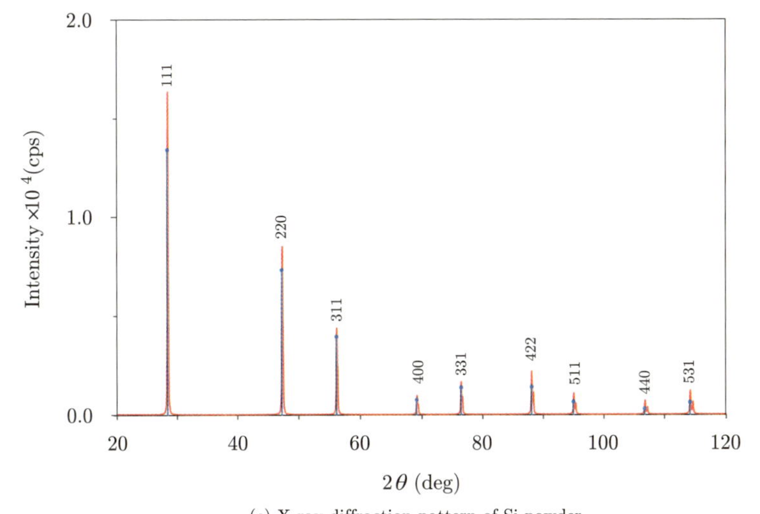

(c) X-ray diffraction pattern of Si powder

Figure C.15 The diamond structure and X-ray diffraction pattern of Si powder. A very sharp diffraction pattern in (c) is one of the characteristics of diffraction line of Si.

the unit cell.

0, 0, 0	1/2, 1/2, 0	1/2, 0, 1/2	0, 1/2, 1/2
1/4, 1/4, 1/4	3/4, 3/4, 1/4	3/4, 1/4, 3/4	1/4, 3/4, 3/4

The difference from the zinc blende structure, therefore, is in the states of their outer electrons. In the ionic crystal, the electron distributions of both types of ions are spherically symmetric in the way that the cation provides one electron to the anion, and the anion accepts it from the cation. In covalent crystal, on the other hand, two electrons released from each atom are localized between the closest atoms and contribute to the tight binding of these two atoms by being shared by them. As an illustration, the bonding in the diamond structure is formed by sharing four electrons in the s- and p-orbital of the L-shell with four adjacent carbons. These electrons form the sp^3 **hybrid orbital** along the four carbon atoms so that its distribution shows anisotropic. The existence of the bonding electrons has been confirmed through the crystal structure analysis of silicon using by X-rays[2], [16]. The lattice parameter of the diamond is 3.56 Å. The lattice parameters of silicon and germanium are 5.43 Å and 5.65 Å, respectively. The tin (Sn) belongs to the IVb group in the periodic table, and it is known also of the diamond structure. Its structure is stable with the lattice parameter of 6.46 Å below the temperature of about 280 K.

The silicon has been used as a basic material in the semiconductor field from the 1960s and the growth of single-crystal with high quality and large-diameter has been demanded. In connection with those things many researches with respect to this material have been examined by scientists. As a typical example, the lattice parameters of silicon have been determined with high precision; silicon has been now used as a standard sample in the measurement of the lattice parameter by using the powder X-ray diffraction. The X-ray diffraction pattern of the silicon showed in Figure C.15 (c).

C.3.2 Quartz

A pure SiO_2 crystal is called the **quartz**. The quartz shows a helical structure along the c-axis of the crystal, which is related to the **optical rotatory power** of this crystal. As shown in the Figure C.16 (a), there are two types of quartz crystal: right-handed quartz and left-handed quartz in which optical rotatory power becomes reverse each other. Here, optical rotatory power is referred to as the physical phenomenon in which the deflecting surface of the light is rotated when passing through a crystal.

The quartz belongs to the hexagonal system as easily recognized from its external form. The lattice parameters are $a = b = 4.912(1)$ Å, $c = 5.402(1)$ Å, $\alpha = \beta = 90°$, $\gamma = 120°$. The unit cell contains three molecules (Figure C.16 (b) and (c)). The position of Si is described by the following three sites with thee parameters $x = 0.5302$, $y = 0$, and $z = 1/3$.

$$(x, y, z), \{ -y, x - y, \mathrm{z} + (1/3)\}, \{-x + y, x, z + (2/3)\}$$

As shown in the Figure C.16 (b) and (c), Si atom (blue color) is located at the center of the tetrahedron formed by O atoms (red color). The Si atom and four O atoms are bonded by the covalent bond. The structure is formed by the three dimensional network of the SiO_4 tetrahedrons. This crystal is also known to show the **piezoelectric effect**. The shape of SiO_4 tetrahedron changes by electric field or pressure. The piezoelectric effect is said to be first found in 1811 by quartz.

The quartz powder sample has 5 characteristic peaks around scattering angle 2θ of 68°, as mentioned in the Chapter 3.6. The 5 characteristic peaks are used for the identification of the quartz in the material analysis. In the diffraction pattern of Figure C.16 (d), all the reflections are observed without extinction rule.

If raised the temperature up to 575°C, quartz shows the phase transition to the high-temperature type quartz which still belongs to the hexagonal system. This phase transition is not abrupt. It is gradually changed within the temperature range of 2 °C.

C.3.3 Graphite

In **graphite**, 2 kinds of structures are known. One is a crystal belonging to a hexagonal system and known most widely as a 2H type. The other belongs to a trigonal system (rhombohedral system) with a 3R type. Both

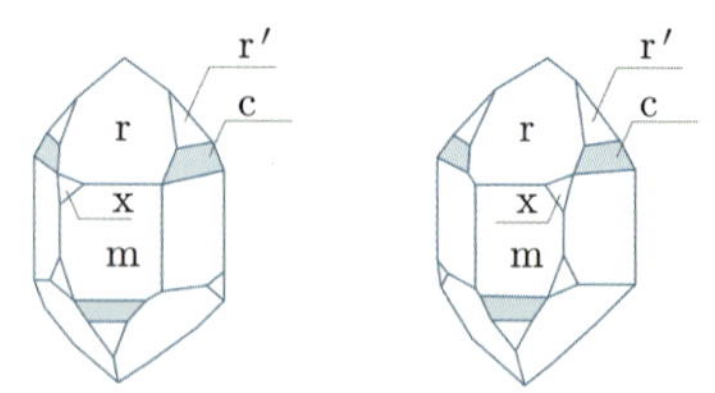

(a) 2-types of quartz (crystal habit)

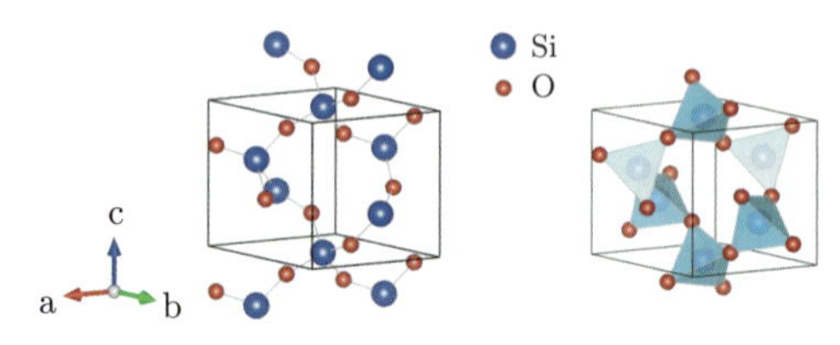

(b) Atomic arrangement in unit cell. (c) SiO_4 tetrahedron

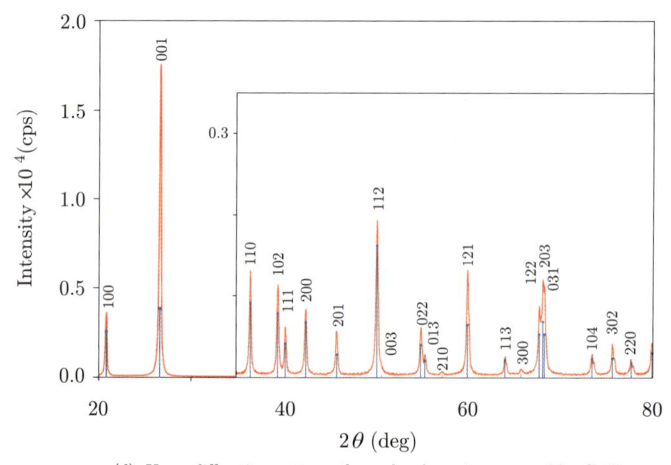

(d) X-ray diffraction pattern of powdered quartz measured by CuKα.

Figure C.16 The outer shape, crystal structure, and X-ray powder diffraction of the quartz.

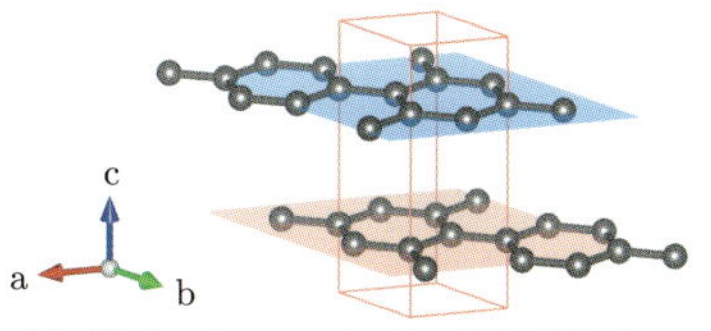

(a) 2H structure, and unit cell (red line).

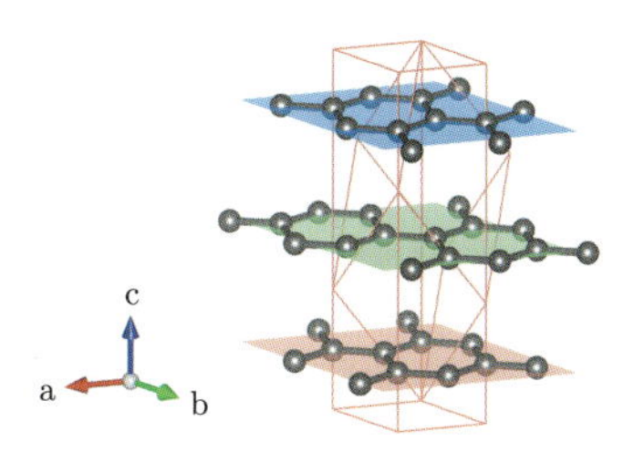

(b) 3R structure, and unit cell (red line).

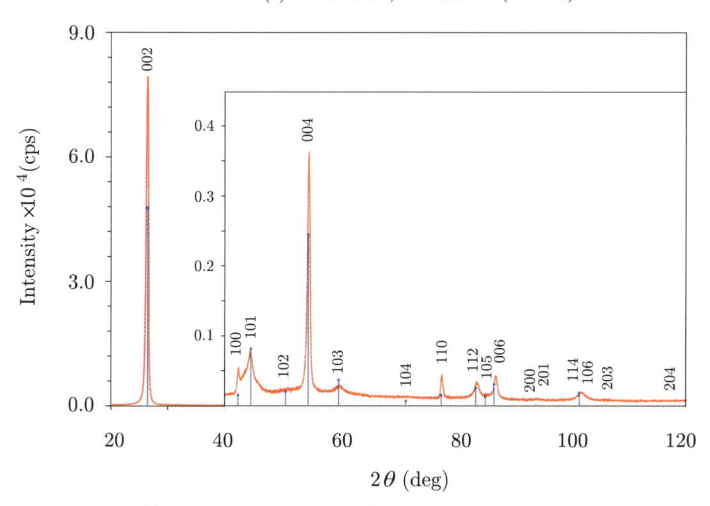

(c) X-ray diffraction pattern of graphite powder measured by CuKα.

Figure C.17 Crystal structure and X-ray diffraction pattern of graphite.

structures are shown in the Figure C.17 (a) and (b). The carbon atoms are located at the hexagonal lattice points in the periodic honeycomb structure and form an atomic plane. Both crystals are a layered structure in which the atomic planes are stacked with a lateral offset. The unit cells were shown by the red line in the figure.

The lattice parameters of 2H structure which belong to the hexagonal are $a = 2.456$ Å and $c = 6.696$ Å and the unit cell contains 4 carbon atoms. The carbon atoms are located at (0, 0, 0) and (1/3, 2/3, 0) in the plane of $z = 0$, and at (0, 0, 1/2) and (2/3, 1/3, 1/2) in the plane of $z = 1/2$. The carbon atom is surrounded by three nearest neighbor atoms in a plane. The distance between carbon atoms is 1.42 Å, which is longer than that of carbon atoms in the benzene but shorter than the distance of carbon atoms in diamond. Therefore, it can be said that the atomic sheet of carbon is constituted by a covalent bonding between carbon atoms. The distance between the two atomic planes is 3.35 Å which is rather long, indicating that atomic planes are bonded by a weak Van der Waals force. It is considered to be the reason why the graphite is a slippery and why powdered graphite is smooth and easily bend.

An existence of the polymorph in the graphite is considered to be reasonable from such a bonding characteristic. The melting point is 3550 °C, which is high compared with that of other elements. This material shows also outstanding anisotropy in thermal conductivity, elasticity, susceptibility, and thermal expansion rate etc. As for electronic structure between carbon atoms in the plane, the atoms are bonded by σ-bond of $(sp)^3$ orbitals. One extra electron of the outer shell is an unpaired π electron and contributes to the conduction, showing the same electrical conductivity with the metal. In contrast, the electrical conductivity along the c-axis is poor by about 4 orders of magnitude. As the result, the electrical conductivity of graphite shows considerable anisotropy. It can be understood that the physical properties of this material is very anisotropic, from the fact that it is of such a layered structure.

If hydrocarbon placed on a plate, which was kept at a high temperature, was decomposed thermally under compressing at high pressure, well-oriented graphite layers grow. It is called the **pyrolytic graphite**. The single crystal is difficult to grow up normally because of weak interlayer bonding. The pyrolytic graphite is, however, conveniently used as the crystal for X-ray spectroscopy, since (0 0 2) plane of the graphite has a large lattice spacing. The Figure C.17 (c) shows the diffraction pattern of the pyrolytic graphite. The (0 0 2) reflection has remarkably high intensity whereas others do not. In addition, the FWHM of the reflection is wider than those of other crystals, which is attributed to small crystallites with defects.

By carbonization of carbon-based polymer material at a high temperature, **glassy carbon** is formed. The color is black. This material is formed by 3-dimensional disorder network linked by carbon atoms so that it is not crystal. The glassy carbon is known as the outstanding material with properties of heat-resistant and chemical resistance.

C.4 Crystal structures of molecular crystals

Generally, there are crystals called molecular crystal. A difference from the other crystals, which have so far been described, is its cohesive mechanism. The molecular crystal is formed by a weak Van der Waals force which are seen between molecules. The molecular crystal has therefore the characteristic of low melting point and low latent heat of melting, in comparison with the other crystals. If a very few difference seen in the structure is ignored, the structure analysis of crystal with X-rays becomes a clue finding the molecular structure. Currently, X-ray diffraction is also used to examine the structure of organic molecule, the structural analysis of the most thriving protein is a good example. However, because the crystal structure depends on the shape and size of the molecule, the most of molecular crystal systems exist with low symmetry, and good quality crystals may be also very few.

Here, let's take up only very simple molecular crystals. First of all, inert gas can be cited such as He, Ne, Ar, Kr, and Xe. At room temperature and under normal pressure, there are diatomic molecules in the gas state, such as H_2, N_2, O_2, Cl_2, Br_2, and CO_2. Slightly more complicated are the hydrogenated compounds of light elements such as HF, HCl, HBr, H_2O, H_2S, H_2Se, NH_3, PH_3 and CH_4. These materials have a low melting temperature, and their structure near the melting point is quite unique and interesting. The structure of Ne, A, Kr, and Xe is FCC, whereas the structure of He is the HCP structure. As you know, the charge density distribution of these inert elements will be regarded as spherically symmetric. In view of this fact, it is understandable that the structure of these elements is the FCC. However, it is hard to know why only He takes the HCP structure.

H_2 as a representative of diatomic molecules shows the HCP structure except extremely low temperature. The form of diatomic molecule is of "cocoon" type that silkworm creates. The molecules are known to be rotating almost freely around the center of gravity even if crystallized. Because the rotation is suppressed at low temperature, it is an interesting subject to investigate what structure does H_2 appears at the low temperature.

Among the hydrogen compounds of light elements, especially HF, H_2O, and NH_3, each has a unique structure, respectively. Thus, they will be described in the next by taking a new section. Let us consider the reason why the crystal structure of H_2 is FCC near the melting point. It is considered that the rotation energy of H_2 molecule is smaller than the heat energy k_BT near the melting point so that almost free rotation is allowed to H_2 molecule. However, when temperature decreases, the interaction between molecules seems to become significant with respect to the rotation energy, since the molecule has multipole moment. This situation may lead to phase transition to complicated peculiar structure. Other molecules hydrogenated also form very interesting structure. Even if it crystallized, the molecules behave like H_2 molecule near the melting temperature, because the energy of the molecular rotation is also small compared with the thermal energy $k_B T$. These are cumbersome gases and it does not solidify unless it is cooled down to a very much lower temperature, any experiment is not easy to perform. For that, although several studies have been examined quite a long time ago, research after those seems to be not so advance[18].

C.5 Crystal structures of hydrogen-bonded crystals

There are crystals of which constituent atoms or molecules are bound by hydrogen located between the two atoms. They make a structure in the form of X–H–X or X–H–Y, where X and Y represent the different light elements which are easy to become the anion such as fluorine (F), oxygen (O), and nitrogen (N). In the structure, the hydrogen releases an electron and becomes proton p^+, the released electron is taken over to the X and the Y atom. As a result, the X and the Y atom become the anion and attract one positive proton p^+ by coulomb force. It may be said that the structure is ionic bonding locally formed by the minus-plus-minus. Such an inter-atomic bonding property called the **hydrogen bonding**.

Although the bonding property is found in one molecule, it contributes also to cohesive mechanism between molecules. The cohesive force is directional and is stronger than Van der Waals forces. Thus, when molecules crystallized, its structure is often observed as deformed from that seen in the gas phase. The hydrogen fluoride (HF)[26], the water (H_2O), and the ammonia (NH_3)[27] molecule is crystallized with such a bonding. The crystal shows, therefore, very unique structure.

C.5.1 Ice

The ice crystal has been known as a representative one with the hydrogen bonding. The ice is obtained from pure water when it freezes at 0° under atmospheric pressure. It is called the Ih phase which takes the ZnS type structure of hexagonal system. Thus, the oxygen atoms occupy the positions of Zn and S. Each oxygen atom is surrounded by 4 neighbor oxygen atoms forming the tetrahedral configuration. Hydrogen atom is situated between two oxygen atoms, although it is closer to one of the oxygen atoms as represented in a form O-H—O. However, the hydrogen moves closer to the other oxygen at another moment, as shown by O—H-O. The oxygen atom becomes a divalent anion by depriving an electron each from two hydrogen atoms, and a hydrogen atom is considered to become a bare proton. So, the proton having a positive charge oscillates back and forth between two divalent oxygen anions. From this viewpoint, the proton does not belong to either of the two oxygen anions. The time average structure is observed as O - p/2−p/2 - O, representing that the proton with the existence probability of half is situated at two positions between the two oxygen ions. The structural model shown is called the half hydrogen bonding. The position of the proton has been confirmed by the Fourier analysis of the experiment data obtained from neutron diffraction[19] and electron diffraction[20], although it is difficult to determine the position of the proton by X-rays. The bonding between the oxygen atoms is kept through the hydrogen, hence the structure is different from that of the molecular crystal bonded by Van der Waals force.

By varying the temperature and the pressure, the structure of ice changes dramatically. The reported crystal structures are listed in Table C.8. Other

Table C.8 8 crystalline states of ice.
Although IV and IX has been reported, they are metastable phases.

Ice	Crystal system	Density (g/cm^3)	Ice	Crystal system	Density (g/cm^3)
I_h	Hexagonal	0.92	V	Monoclinic	1.23
I_c	Cubic	0.93	VI	Tetragonal	1.31
II	Trigonal (rhombohedron)	1.17	VII	Cubic	-1.50
III	Tetragonal	1.16	VIII	Cubic	-1.66

phases, except I and II, are the high pressure phases that appear at pressures exceeding 2×10^8 Pa. In the table, I_c is a special phase; it is a frozen fog, may be said to be frost, existed on the cooled metal in vacuum. The crystal structure is the same as cubic ZnS, although a similar half hydrogen bonding to that of I_h exists. The position of hydrogen atom has been clarified, as mentioned above, by the experiment of electronic diffraction.

C.5.2 Hydrogen bonding crystals except ice

Examples of the crystal other than the ice that are bound by hydrogen bonds are hydrogen fluoride (HF) and ammonia (NH_3). In addition, the lowest temperature phase of hydrogen sulfide (H_2S) shows also a structure bonded by the hydrogen bonding, according to electronic diffraction experiment[21].

As a crystal showing ferroelectric crystal in which hydrogen bonding is involving, potassium dihydrogen phosphate (KH_2PO_4 it abbreviates to KDP) is represented here. The KDP is the tetragonal structure of the lattice parameter $a = 7.452\,Å$ and $c = 6.974\,Å$ at room temperature. It shows paraelectricity. At a temperature $T_c = 123\,K$ ($-150\,°C$), the structure changes to the orthorhombic, and the system shows ferroelectricity[22]. The phase transition is of the order-disorder type in which hydrogen bonding participates.

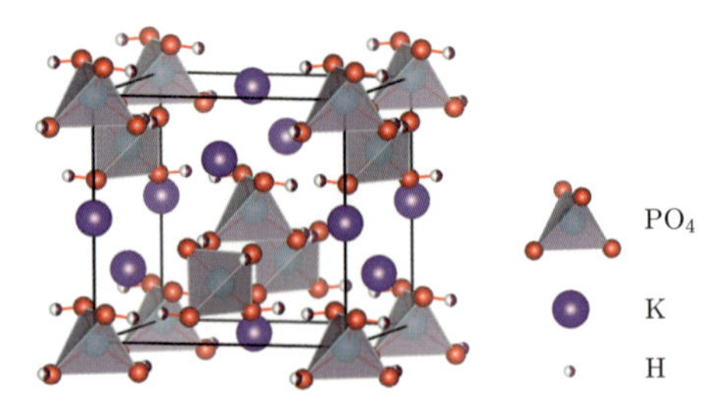

Figure C.18 Crystal structure of high temperature phase of KH_2PO_4 (KDP).

The structure of the high-temperature phase consists of a positive potassium ions K^+ and a tetrahedron $(H_2PO_4)^-$ of monovalent anion. Oxygen at the four corners of the PO_4 tetrahedron are connected with the oxygen from the adjacent (PO_4) by hydrogen bonding represented by O-H $\cdots$ O. Here, all the hydrogen bondings are almost parallel to *ab*-plane as seen in Figure C.18. Two hydrogens belong to one (PO_4) tetrahedron, but they are not necessarily localized between the two O atoms, located in the irregular state, in the paraelectric phase. In the structure of the low-temperature phase, spatial arrangement of O atoms is almost the same as those of the high-temperature phase, K and P shift by only slight amount 0.04 Å and 0.08 Å, respectively, along the *c*-axis direction, and the crystal is deformed. On the other hand, the hydrogen atom is localized one of the O atoms as $(H_2PO_4)^-$. In order to clarify the hydrogen positions as well as the structural change, the neutron diffraction technique is utilized with KD_2PO_4 (called DKDP, $T_c = 213\,K$) in which hydrogen of KDP is replaced by deuterium.

The KDP is known as a typical substance showing order-disorder type and compared with $BaTiO_3$ of displacive type in the ferroelectric phase transition.

C.6 Other crystals

As novel materials, there are high-temperature superconducting materials, magnetic materials, asbestos-related materials, cement mineral-related special materials, and pharmaceutical compounds. In addition, the crystal materials can be classified by inorganic and organic materials. The classification itself may be thought in various ways. It will be your future problem from which you came here read up. Select the material that you are interested in and examine the relevant crystal until you are sufficient. It is recommended to create a database of our own.

C.7 Other factors including cohesive force

When crystalline material is heated up to melting point under normal pressure, the material absorbs a latent heat of melting to change into liquid state. Upon further heating the liquid to the boiling point, the liquid absorbs a latent heat of vaporization to evaporate, and is resolved into atoms or molecules. Those changes of the states are called the phase transitions of solid-liquid-gas. When the melting temperature is significantly high value and also the latent heat of melting is large, it is recognized that the material is of a strong cohesive force and in a stable structure. On the basis of large or small of those value, as shown above, the materials can be classified into five groups having different binding energy.

The strongest chemical bond is covalent bond, showing the energy of bonding to be 170 – 290 kcal/mol. The next strongest bonding is seen in the ionic crystal of which bonding energy is in 120 – 250 kcal/mol. The cohesive energy of metallic bonding is in 25 – 95 kcal/mol. Hydrogen bonding and Van der Waals bonding are in 7 – 12 kcal/mol and 2.0 – 2.5 kcal/mol, respectively. It should be noted here that the bonding energies of these crystallites are small value by more than one digit. What difference in cohesive force is reflected in the crystal structure? Unfortunately, the relationship between cohesive force and crystal structure remains unclear, as mentioned at the beginning of this appendix.

Note that there is always a certain temperature range in which the structure is stable. A certain crystal structure, which has been stable in low temperature, changes into a different structure at an elevated temperature. There are a lot of such materials. As mentioned previously, the change of the structure is called the structural phase transition in solid. The question then arises: "why the crystal performs phase transition to a different crystal structure at a given temperature." The answer given by statistical thermodynamics is that the phase change, which happens in a solid state under constant temperature and pressure, is accomplished by the Gibbs free energy given by $G = U + PV - TS$. Where, U, P, T and S are the internal energy, the pressure, the temperature and the entropy, respectively. It should be noted that the phase change is not accomplished only by the internal energy U representing the magnitude of cohesive force of crystal, but also by change of orientation, thermal vibration, and rotation of atoms and/or molecules, which are kind of the indicator for degree of entropy. All of those freedoms are dependent of crystal structure. If atoms and/or molecules are not able to freely move in a given crystal, the entropy term ST is not sufficiently large even at the high temperature. The Gibbs free energy G will not be low because of the low entropy term ST, although the structure is of strong cohesive force. However, another structure, which is of small cohesive force but large ST, will show more low free energy of G. Thus, this structure becomes more stable and will supersede the original structure. This explains why a phase transition occurs at a given temperature.

Above argument might be explained more clearly by using a graph of free energy G versus temperature T. The Gibbs free energy is given by

$G = U + PV - TS$. Thus, G is the function of temperature T and pressure. Here, U itself is not temperature dependent and gives the amount of the binding energy represented by negative quantity. Although the volume might depend very slightly on temperature, we regard PV as a constant in this discussion. We thus ignore this term. As for the entropy term ST, the entropy S is strongly dependent on the crystal structure. Let us denote as S_1 and S_2 as the entropies of the initial ($j = 1$) and final ($j = 2$) structure in the phase transition, respectively. In such a case, G is obtained by a straight line as shown in Figure C.19. In this figure, two straight lines $G_1 = U_1 - S_1T$ and $G_2 = U_2 - S_2T$ are shown for two structures. Since $|U_1| \gg |U_2|$, but $S_1 \ll S_2$, we can see that the free energy G_1 of the structure $j = 1$ is lower than the G_2 of the structure $j = 2$, so the structure $j = 1$ is stable in the temperature lower than T_c. However, reverse situation occurs in the temperature above the T_c, so that the structure $j = 2$ becomes stable. This result indicates that the phase transition occurs at the cross point of two straight lines.

There is one more topic to be noted. It is hard to distinguish the cohesive forces of ionic crystal from those of covalent crystals by the structure of

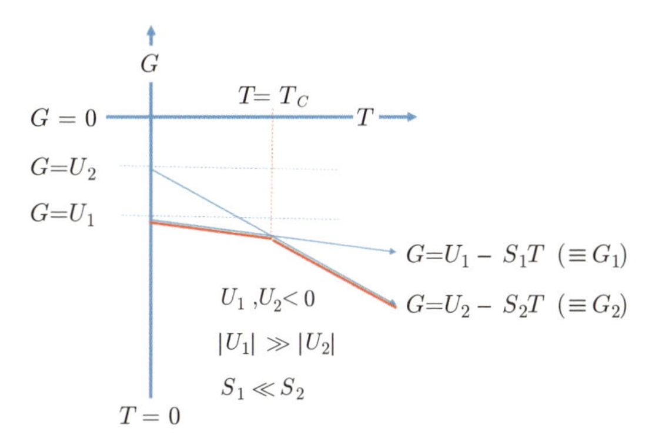

Figure C.19 Gibbs free energy vs. temperature.

their crystals even if represented by atomic positions with thermal parameters. Recently, a lot of researches are attempting to find a difference of such cohesive forces by using X-ray diffraction, which precisely visualizes electron density distribution, with the help of theoretical calculation. The electron density distribution could be reproduced as spherically symmetric around an ion in ionic crystal and anisotropic shape around the atom in covalent crystal. It is now possible for the study to achieve by using a powder X-ray diffraction method with MEM analysis, although the details of this approach are not presented here.

Annotations

1) Since the figure is a projection view depicting here, it might be difficult to understand the three-dimensional view. Draw it by yourself by creating a model and verify the number of nearest neighbor atoms.

2) The first chapter refers to α-Fe, which infers indirectly the existence of β-Fe. Iron has a magnetic property at room temperature. We call this state "ferromagnetic". When it comes to 1043 K (770 °C) by raising the temperature, the magnetic property will be lost. It refers to the state of paramagnetic. However, the structure does not change and stay in the structure of BCC. Originally, the paramagnetic state was not well known, so this state was called β-iron. In a dictionary of East-West physical science, the word "ferrite" is used as a common name of materials with magnetic property, including iron oxids.

3) "nm" is used in MKS units. In crystallography, $1/10\,\mathrm{nm} = 1\,\text{Å}$ is used as unit. The unit is convenient for showing the spacing of crystal lattice as well as the wavelength of incident X-rays and neutron beam. In this Textbook, "Å" is used for the unit.

4) Under the same temperature and the same pressure, there are several materials which show a different crystal structure; it is called as polymorph as mentioned already in early chapter. The physical property is different because of the different crystal structure. A typical example of simple structure formed of one element will be carbon. The graphite and the diamond are in a relation of the polymorph. Recently, graphite polymorphs have been found such as fullerene and carbon nano-tube. These materials have attracted much attention.

References

〔1〕ICDD：The Powder Diffraction File & Related Products, sales catalog 2013-2014
〔2〕M. Sakata: *Butsuri* **48** (2) (1993) 78
M. Sakata, E. Nishibori, M. Takata: *Netsu Sokutei* **31** (1) (2004) 29
〔3〕W. H. Zachariasen: *J. Am Chem. Soc.* **54** (1932) 3841 and also see textbook given by C. Kittel
〔4〕For example: W. H. Zachariasen: Theory of X-ray Diffraction in Crystal (1945) John Wiley and Sons Inc. Chapter III Sec. 8-14, 111-155 or N. Kato: "X-sen kaisetsu to kouzou hyouka" (1978) Asakura Co. Chapter 5, p.43-
〔5〕K. Ohshima, J. Harada and N. Sakabe: X-ray Instrumentation for the Photon Factory, edited by S. Hosoya, Y. Iitaka and H. Hashizume (1986) 35
〔6〕P. Scherrer: Goettinger Nachr. (1918) p.98, S. Miyake: "X-sen no kaisetsu" (1969) Asakura Co., B.E. Warren: X-Ray Diffraction (1969) Addison-Wesley Pub.Co.
〔7〕W. H. Hall: *Proc. Phys. Soc.*, **A62** (1949) 741, *J. Inst. Met.* **75** (1950) 1127
〔8〕W. A. Dollase: *J. Appl. Cryst.* **19**, (1986) 267, M. Ahtee, M. Nurmela, P. Suoritti and M. Jaervinen: *J. Appl. Cryst.* **22**, (1989) 261
〔9〕International Table for X-ray Crystallography Vol. III (1962), Vol. C (1992) & (1999), Vol.A (2005)
〔10〕J. Harada: *Butsuri* **18** (1963) 36
〔11〕ICDD File (2009): www.icdd.com
〔12〕S. Nagakura, *et al*.: Rikagaku jiten 5th Edi. (1999) Iwanami Pub. Co.
〔13〕F. Yona and G. Shirane: Ferroelectric Crystals (1962) Pergamon Press, London
E. Fatuzzo and W. J. Mrz: Ferroelectricity (1967) North-Holland Publishing Co.
〔14〕J. Harada: *J. Cryst. Soc. Japan* **20** (1978) 1
〔15〕Londolt-Boernstein New Series, Group III, Ferroelectrics and Related Substances, *a* oxides **16** (1981) and **28** (1990) Sprng-Verlag Berlin
〔16〕H. Hattori, H. Kuriyama, T. Katagawa and N. Kato: *J. Phys. Soc. Jpn.* **20** (1965)
〔17〕I. Nitta: *J. Cryst. Soc. Japan* **1** (1959) 1
〔18〕S. W. Peterson and H. Levy: *Acta. Cryst.* **10** (1957) 70
〔19〕K. Shimaoka: *J. Phys. Soc. Japan* **15** (1960) 106
〔20〕J. Harada and N. Kitamura: *J. Phys. Soc. Japan* **19** (1964) 328
〔21〕J. Harada: *J. Cryst. Soc. Japan* **45** (2003) 306
〔22〕T. Ito: *Nature* **164** (1949) 755, J. W. Visser: *J. Appl. Cryst.* **2** (1969) 89, D. Louer and N. Louer: *J. Appl. Cryst.* **5** (1972) 271

〔23〕 H. M. Rietveld: *J. Appl. Cryst.* **2** (1969) 65
F. Izumi: *J. Cryst. Soc. Japan* **27** (1985) 23
F. Izumi: " Atarasii funmatsu kaisetsu hou " (1992) Chap. IV, p.67-, edited by The Crystallographic Society of Japan
〔24〕 K. Ohshima, S. Yatsuya and J. Harada: *J. Phys. Soc. Japan* **50** (1981) 861
〔25〕 M. Atoji and Lipscomb: *Acta Cryst.* **7** (1954) 173
〔26〕 I. Olovsson and D. H. Templeton: *Acta Cryst.* **12** (1959) 832
〔27〕 M. Sakata and J. Harada: *J. Cryst. Soc. Japan* **22** (1980) 387

Reference Books

L. Bragg: The Crystalline State Vol.I-A General Survey- (1955) G. Bell and Sons Ltd..

J. C. Slater and N. H. Frank: Introduction to Theoretical Physics (1933) McGraw-Hill Book Co. Inc. New York and London.

J. C. Slater: Introduction to Chemical Physics (1939) McGraw-Hill Book Co. Inc. New York and London.

W. H. Zachariasen: Theory of X-ray Diffraction in Crystals (1944) Dover Publications, Inc. New York.

B. E. Warren: X-ray Diffraction (1969) Addison-Wesley Publishing Company, Massachusetts.

C. Kittel: Introduction to Solid State Physics, 2nd (1956), 4th (1971) and 5th edition (1976) John Wiley & Sons, Inc. New York.

Azarof/ Kaplow / Kato/ Weiss/ Young: X-ray Diffraction (1974) McGraw-Hill Book Co. Inc. New York and London.

J. M. Cowley: Diffraction Physics (1975) North-Holland publishing Co. Amsterdam and Oxford.

A. Guiener: "X-sen Kesshougaku no Rironn to Jissai" translated by K. Kora, S. Hosoya, K. Doi and K. Niizeki (1967) Rigaku Co..

S. Miyake: "X-sen no Kaisetsu" (1969) Asakura Pub. Co. Ltd..

T. Sakurai: "X-sen Kesshou Kaiseki" (1970) Shokabo Co. Ltd..

Y. Saito: "Kagaku Kesshougaku" - Ekkususen Kesshou Kaiseki no Kiso - (1975) Kyoritu Pub. Co..

N. Kato: "Kaisetu to Sanran" (1978) Asakura Pub. Co. Ltd..

Y. Kainuma: "Kannshou oyobi Kanshousei" - One point Physics - (1981) Kyoritu Pub. Co..

K. Kora (Chief Editor): "X-sen Kaisetu" (1988) Kyoritu Pub. Co..

N. Kato: "X-sen Kaisetu to Kouzo Hyouka" (1995) Asakura Pub. Co. Ltd..

The Crystallographic Society of Japan: "Atarashii Funnmatu Kaisetu-hou" - Text book for training course - (1992)

The Crystallographic Society of Japan, Editorial board of "Handbook of Crystal Analysis", J. Harada (Chief Editor) "Kessho Kaiseki Handbook" (1999) Kyoritu Pub. Co..

Rigaku Co.: "X-sen Kaisetu Handbook" (1999).

Index

Author

Jimpei HARADA
Professor Emeritus, Nagoya University, Nagoya
(Former a Senior Vice President of Rigaku Corporation
and the Director of X-ray Research Laboratory, Tokyo)

Powder X-ray Diffractometry in the Analysis of Materials
－Utilization of MiniFlex－

2016年 9月 30日 発 行

著作者 原 田 仁 平

発行所 丸善プラネット株式会社
〒101-0051 東京都千代田区神田神保町二丁目17番
電 話(03)3512-8516
http://planet.maruzen.co.jp/

発売所 丸善出版株式会社
〒101-0051 東京都千代田区神田神保町二丁目17番
電 話(03)3512-3256
http://pub.maruzen.co.jp/

組版・印刷・製本 三美印刷株式会社

ISBN978-4-86345-302-9 C3042